Managing the Asian Century

Series editor

Purnendu Mandal, College of Business, Lamar University, Beaumont, TX, USA

Managing the Asian Century provides a platform for scholastic discussions and stresses the need for a holistic framework to understand Asia as an emerging economic global powerhouse. Books published in this series cover Asia-centric topics in economics, production, marketing, finance, entrepreneurship, education, culture, technology, as well as other areas of importance to Asian economics. The series will publish edited volumes based on papers submitted to international and regional conferences that focus on specific Asia-Pacific themes, such as investment in education, women's rights, entrepreneurship, climate change, wage inequality, challenges in governance, and corruption. Books in this series are of keen interest to researchers and policy planners around the world and will be used by universities for graduate and doctoral level studies.

More information about this series at http://www.springer.com/series/13579

Pradip Kumar Ray · Jhareswar Maiti
Editors

Ergonomic Design of Products and Worksystems—21st Century Perspectives of Asia

Springer

Editors
Pradip Kumar Ray
Indian Institute of Technology Kharagpur
Kharagpur
India

Jhareswar Maiti
Indian Institute of Technology Kharagpur
Kharagpur
India

ISSN 2364-5857 ISSN 2364-5865 (electronic)
Managing the Asian Century
ISBN 978-981-10-5456-3 ISBN 978-981-10-5457-0 (eBook)
https://doi.org/10.1007/978-981-10-5457-0

Library of Congress Control Number: 2017945236

Printed on acid-free paper

This Springer imprint is published by Springer Nature
The registered company is Springer Nature Singapore Pte Ltd.
The registered company address is: 152 Beach Road, #21-01/04 Gateway East, Singapore 189721, Singapore

Preface

Design is the central activity of the engineering profession. In today's context, as human–machine interface design or simply interface design has become a core issue, it is essential that design engineers and technologists need to have the knowledge of ergonomics and human factors engineering. Ergonomic design and analysis has emerged as one of the key areas of study and research over the past several decades in order to make the worksystems ergonomically designed, operationally safe, and productive. The book addresses the major issues, principles, and applications related to ergonomic design of products and worksystems.

The book is a comprehensive text on the Proceedings of the International Conference on Management of Ergonomic Design, Industrial Safety and Healthcare Systems (MESH-2016) held at Indian Institute of Technology Kharagpur, India, during December 20 to 23, 2016. It is self-contained and the subject matter is presented in an organized and systematic manner. Each chapter in this book consists of a concise but thorough fundamental statement of introduction, methodology, and findings followed by results and discussion. The concepts and case studies as presented in the book pertain specifically to the situations and conditions prevailing in Asian countries and their perspectives.

This book is quite appropriate primarily for graduate students, researchers, and industrial professionals. It can also be adapted for a short professional course on the subject matter. It is expected that design and research engineers will be able to draw upon the book in selecting and developing mathematical models for analytical and design purposes. Also, practicing engineers and managers will be highly benefited by learning the basic principles and concepts of ergonomic design as discussed and illustrated in the book.

Kharagpur, India

Pradip Kumar Ray
Jhareswar Maiti

Contents

1 **Ergonomic Performance and Evaluation of Worksystem: A Few Applications** 1
P.K. Ray, V.K. Tewari and Esha Saha

2 **Decision Tree for Manual Material Handling Tasks Using WEKA** 13
R. Rajesh, J. Maiti and M. Reena

3 **Interventions for Combined Manual Material Handling Tasks Based on Biomechanical and Physiological Responses** 25
R. Rajesh and J. Maiti

4 **Computer-Assisted Industrial Ergonomics: A Review** 37
Harish Kumar Sharma, Piyush Singhal and Pankaj Sonia

5 **Ergonomic Analysis of Work-Related Musculoskeletal Disorders for Radial Drilling Machine Operators** 49
P. Ashok, G. Madhan Mohan and B. Siva Sankaran

6 **Ergonomic Evaluation of Bar Benders in Construction Industry** . . . 57
S. Drisya, B.B. Das and Ratri Parida

7 **Formulation of Mathematical Model to Investigate Magnitude of Musculoskeletal Pains Among Cotton Spinning Operators** 69
Dhananjay Ikhar and Vishwas Deshpande

8 **Ergonomic Study on Work Postures of Sewing Machine Operators in Government Industry: A Case in Lean Environment Garment Industry** 83
T. Sakthi Nagaraj and R. Jeyapaul

9 **Ergonomic Design and Development of Material Handling Equipment for Pantry Car Personnel** 103
Abhijeet Arvind Raut and Jyotiraman De

10 Study on Work-Related Musculoskeletal Disorders among Coir Industry Workers in the State of Kerala, India 117
M. Satheeshkumar and K. Krishnakumar

11 Musculoskeletal Disorders and Workplace Analysis of Assembly Section in a Submersible Pump Industry: An EMG-Based Approach 131
P. Ashok, G. Madhan Mohan and K. Nitish

12 QFD-Based Ergonomic Design of Drafting Table for Engineering Students: A Case Study 139
K.G. Durga Prasad, M.V. Prasad, Ch. Hima Gireesh and V.V.V.N.K. Chaitanya

13 Ergonomics for Hand Screen Printing Workers: Cognitive Perception 155
S Shankar, J Karthick and R Naveenkumar

14 Effectiveness of Classroom Interactions in Engineering Colleges 163
L Princy, S Swapnesh, R Rajesh and K Hridya

Author Index 177

Subject Index 179

Chapter 1
Ergonomic Performance and Evaluation of Worksystem: A Few Applications

P.K. Ray, V.K. Tewari and Esha Saha

Abstract Application of ergonomic principles provides a standardized approach for the analysis of a worksystem with emphasis on evaluation of interaction between human(s), machine(s), and environment, as a first step toward exploring the enormous potential and concepts of ergonomics at workplaces. The objectives of the study are manifold: to identify and characterize the ergonomic variables for a given worksystem with regard to work efficiency, operator safety, and working conditions, and to design a comprehensive ergonomic performance indicator (EPI) for quantitative determination of the ergonomic status and maturity of a given worksystem. The study consists of three phases: preparation and data collection, detailed structuring and validation of EPI model. Identification of ergonomic performance factors, development of interaction matrix, design of assessment tool, and testing and validation of assessment tool (EPI) in varied situations are the major steps in these phases. The case study discusses in detail the EPI model and its applications.

Keywords EPI structure · Worksystems · Principal parameters · Normalized total rating

1.1 Introduction

The principles of ergonomics may be applied to the study and design of the components of any worksystem involving human(s) and machine(s) embedded in an environment, and as such its areas of application are not limited by a particular

P.K. Ray (✉) · E. Saha
Department of Industrial Engineering and Management, Indian Institute of Technology Kharagpur, Kharagpur 721302, West Bengal, India
e-mail: pkr@vgsom.iitkgp.ernet.in

V.K. Tewari
Department of Agricultural and Food Engineering, Indian Institute of Technology Kharagpur, Kharagpur 721302, West Bengal, India

P.K. Ray and J. Maiti (eds.), *Ergonomic Design of Products and Worksystems—21st Century Perspectives of Asia*, Managing the Asian Century,
https://doi.org/10.1007/978-981-10-5457-0_1

technology or by the scale of the system. In essence, application of these principles provides a standardized approach for the analysis of any worksystem with emphasis on consideration of interaction between human(s), machine(s), and environment. In this context, Tata Steel, Jamshedpur has been earnestly striving to apply the ergonomic principles, at its worksystems at different levels in order to ensure safety, health, convenience, and comfort of the personnel at their workplaces at an acceptable level of productivity and reliability of the overall system.

Application of ergonomic principles provides a standardized approach for the analysis of a worksystem with emphasis on evaluation of interaction between human(s), machine(s), and environment, as a first step toward exploring the enormous potential and concepts of ergonomics at workplaces, the management of Tata Steel, the largest steel-making company of India in the private sector, has taken steps to institutionalize the process of implementing ergonomics as a whole and felt the need to develop a framework to determine the levels of ergonomic performance at its different workplaces. As the idea of developing an ergonomic measurement system is quite unique, Tata Steel ventures into the design of a comprehensive approach for determining the ergonomic status of the worksystem.

The factors of performance and/or operations where deficiencies or nonconformances occur should be identified and assessed on a regular basis to improve the performance, productivity, and reliability of any unit of analysis, and application of the concept of "remedial" ergonomics in many areas, operations, and factors of production may lead to substantial improvement in overall system performance. An assessment tool for determination of the status (level of ergonomic maturity) and level of ergonomic intervention to be employed in a given worksystem is all that is needed (Ray and Tewari 2012). The objectives of the project are (i) to identify and characterize the ergonomic variables for a given worksystem with regard to work efficiency, operator safety, and working conditions, (ii) to design a comprehensive ergonomic performance indicator (EPI) for quantitative determination of the ergonomic status (in terms of design requirements and performance leveling) of a given worksystem or unit of analysis, and (iii) to apply the EPI model to assess the degree of ergonomic maturity of a given worksystem or unit of analysis.

1.2 Designing an Ergonomic Performance Indicator (EPI)

In order to design an ergonomic performance indicator (EPI) for determination of the ergonomic status of a given worksystem, modeling a general framework involving all relevant ergonomic factors is the essential first step. Assessing the degree of ergonomic maturity against each identified factor is the next important step. The ergonomic factors to be considered in almost all situations or worksystems are related to four key aspects of worksystems, viz, human characteristics, physical workspace, physical environment, and organizational factors. While assessing the level of an ergonomic factor, three principal parameters need to be looked into, viz., work efficiency, operator safety, and working condition, each of which needs to be defined and interpreted in the widest possible sense during evaluation. The principal parameters are (i) "Operator Safety" refers to either levels of outputs obtained per unit of time, or optimum time utilization, or minimum error rate in tasks, or efficiency in manual handling, or minimum energy expenditure rate by the person(s) concerned in a given worksystem; (ii) "Work Safety" may be interpreted in terms of potential danger to health associated with the tasks, deteriorating fitness of the individual concerned, possibility of injuries and accidents, and hazards of any other kind, and (iii) "Working Condition" refers to the condition or environment in the workplace and its surroundings made available to the satisfaction or dissatisfaction and comfort or discomfort of the person(s) concerned. In essence, the performance of any worksystem is a reflection of the joint effect of performance of the individual components of the worksystem, viz., "human", "machine", and "environment".

It is reasonable to assume that for a worksystem to be capable of performing at its maximum level, each component must also contribute significantly and equally to the overall performance of the worksystem as a necessary condition for an acceptable worksystem. However, the central focus of ergonomics being the human(s) in a worksystem, the main consideration in the design of EPI is to measure and assess the state and the effect of human(s) in the worksystem, as a sufficient condition for sustained ergonomic performance. Hence, in order to assign values to the base parameters in EPI modeling, the effect on each component needs to be considered.

The three major components of "human", "machine", and "environment" have been assigned equal weightage while assessing the overall performance of a given worksystem. Against a base parameter, a three-level (L-I, L-II, and L-III) performance rating scale is recommended. For L-I, since the machine and the environment have the highest contribution and the human has the least contribution signifying the level of ergonomic maturity of the system, the total contribution of machine and environment is worked out to be 6.6 in a scale of 0–10. For simplicity and convenience of use, a rounded off figure of 6 has been assigned to this level. Following the same logic, other two levels (L-II and L-III), depending on their ergonomic maturity, have been assigned a maximum score of 12 and 18, respectively.

In the overall assessment of any type of worksystem (manufacturing or non-manufacturing, including service or office activities), various kinds of training programmes on safety may be initiated for the working personnel through safety campaign, and introduction of safe working methods. With regard to such conditions, it is proposed to assign an additional scale rating in a scale of 0–10, on the basis of the intensiveness of safety programmes existing in the worksystem.

1.3 Methodology

The important steps of the methodology leading to the design and development of the EPI assessment tool are as follows:

(i) Identification of ergonomic factors

A general framework involving all relevant factors and sub-factors related to human characteristics, physical workspace, physical environment, and organizational factors is required to be developed including the procedures and norms to be followed for a given unit of analysis (worksystem). A structured data collection form, called EPI data record sheet (version 1) and coded as EPI-DRS-1, has been designed to elicit information and individual judgment from the Tata Steel team members on the ergonomic factors as identified by consultants, and to be included in the design of EPI.

(ii) Identification of design and performance factors, a list of factors related to three specific aspects, viz., operator safety, work efficiency, and working condition including functional requirements, if any, is prepared and standardized at this stage. Wherever feasible and desired, these factors are objectively analyzed with respect to their current levels and/or implications; otherwise, they are subjectively assessed.

Initially, to identify all the possible ergonomic factors so as to have adequate breadth (coverage) and depth (intensiveness) in the EPI model, information on several ergonomic aspects in each factor needs to be considered. With an in-depth understanding of the prevailing situations and requirements of Tata Steel workplaces, several aspects representing depth of each factor, are identified by the consultants.

(iii) Development of interaction matrix

At this stage, the interactions (strong or weak) between the ergonomic and design or performance factors to be ascertained for a given unit of analysis were prepared in order to limit the number of factors with which a given worksystem may be assessed to a reasonable size. The guidelines for the selection of appropriate number of factors were established. The rules for determining the relative weights (reflecting importance or criticality of a factor in the presence of other factors, or on its own) were specified at this stage.

(iv) Design of the assessment tool

On completion of the above three steps, a comprehensive framework for (1) determining the ergonomic performance of a worksystem (the basic EPI model), (2) identification of deficient area(s) in relation to ergonomic factor(s), and (3) setting the priority of improvement actions suggested, is established at this stage. Therefore, a structured methodology for measuring ergonomic performance of a worksystem, called EPI data record sheet (version 2) and coded as EPI-DRS-2, is developed by the consultants for circulation and use among the Tata Steel personnel.

(v) Testing and validation of assessment tool (EPI) in varied situations and worksystems

The proposed tool is to be tested for its verification, validation, and applicability in a number of representative situations as specified and identified by the management of Tata Steel during the course of study. Appropriate modification of the assessment tool is required to be made based on actual observations, review of opinion of Tata Steel personnel, and performance evaluation.

(vi) Data collection and analysis

The entire data collection and analysis was completed by the consultants through (i) meetings with the Tata Steel team members and other concerned personnel at regular and planned intervals, (ii) in-house preparation of required documents (EPI-DRS-1, EPI-DRS-2, and interaction matrix), (iii) visits to the selected and representative worksystems and departments at Tisco plant, (iv) verification and validation of the EPI model by hands-on exercises, and (v) discussion on the EPI model.

1.4 The EPI Model

The details of the EPI model designed and developed for Tata Steel, Jamshedpur, are described in three interrelated parts: Part-I, Part-II, and Part-III. Part-I lists the ergonomic factors to be considered for EPI as well as the guidelines for quantitative assessment of base parameters in the design of EPI. Part-II refers to the tables required to be used to compute the EPI score of a given worksystem. There are five tables listed. The first table provides the definition of scale values for leveling and rating of the ergonomic variables considered; second table shows the details of the scale values under different conditions for assessment of visual environment; the third set of tables present the scale values under different conditions for assessment of thermal environment that requires evaluation of radiant temperature, air speed, and relative humidity, separately; the fourth table presents the details of the scale values under specified conditions and jobs for assessment of auditory environment. The data given in these tables are applicable in acclimatized conditions. The fifth table lists the possible EPI grading of a worksystem under consideration as well as status and actions for improvement. Part-III describes the systematic process of determining the EPI score of a given worksystem.

To understand how the EPI score is computed for a worksystem, the features and working of each part is to be known. The details of the features and working of each part are given below.

Part-I is designed to understand and quantitatively assess the importance of base parameters for a worksystem. In order to help define, assess, and quantify a parameter in the most logical and objective way, each ergonomic factor with its scale value is required to be defined for an objective assessment of base parameters. It is opined that the conditions as described in the guidelines are an exhaustive representation of different working conditions and systems at the present level of technology at Tata Steel.

It is recommended that the analyst studies the prevailing conditions against the following factors considered, with regard to key principal parameters viz., work efficiency, operator safety, and working condition, and matches with those given in the guidelines below. Against each factor, three specific rating scales representing acceptable to unacceptable levels are provided.

1. Pace or speed of work under the control of the operator
2. Adequacy of fatigue allowances for jobs
3. Workers away from their workplace during work
4. Occurrence of "human" errors
5. Frequency of lifting of weights
6. Force required to push or pull objects
7. Movements of human body
8. Assessment of visual environment in the work place
9. Engineering Anthropometry
10. Work Posture
11. Assessment of thermal environment in the work place
12. Workers complain about physical environment in their workplaces
13. Tasks resulting in excessive material wastes
14. Repetitive motions/ frequent use of hand tools/ both hands and feet operating/ same posture/information overload/in sufficient time to sense and respond to signals/ physical fitness/ knowledge of training
15. Assessment of Auditory Environment.

In Part-II, each ergonomic variable or base parameter to be considered should be leveled and quantified, on a predefined scale, once the conditions are known to the analyst. Based on the consideration of the ergonomic effect of the specified conditions on the components of the worksystem, each condition is required to be quantified.

For determining the scale values of a specified ergonomic variable, a three-point scale is found to be applicable, feasible, and easily implementable. Each scale value, for which a level number is given, indicates a numeric ergonomic assessment score of a given situation or worksystem, based on the degree of maturity in the "human" component at the existing level of technology (defined in terms of "machine" and "environment" components). The definition of each level is provided.

While rating a given condition, it is quite probable that the degree of maturity of either "machine" or "environment" or both may also be medium or low. Under such conditions, the scale value is required to be suitably modified towards the lower side.

In Part-III, the steps to be followed for obtaining the EPI score of a worksystem are given. These steps are as follows:

Step-1: Select the principal parameter(s) relevant for the worksystem under consideration.
Step-2: Select the base parameter(s) influencing the identified principal parameter (s) in Step-1.
Step-3: Assess the situation against each base parameter considered (as described in Part-I Sect. 5.1), and assign its scale rating (SR).
Step-4: Repeat Step-3 for all other base parameters selected.
Step-5: Compute the sum of scale ratings (SRs) obtained in Step-3 and Step-4.
Step-6: Assess the intensiveness of safety programmes adopted, and assign an appropriate safety awareness rating in a scale of (0–10).
Step-7: Compute the total ratings obtained in Step-5 and Step-6.
Step-8: Compute the normalized total rating **(NTR)** in a scale of (0–100).

The normalized total rating (NTR), as obtained in Step-8 above for a given worksystem may be graded belonging to one of the five specific classes of worksystems given in Table 1.1. By using the EPI model, a worksystem may be of one of the five types: for Class-I or "excellent worksystem", the prevailing work conditions need to be maintained, for Class-II or "very good worksystem" refers to comparatively acceptable work condition and remedial steps wherever required may be initiated, for Class-III or "good worksystem" refers to acceptable work condition with a great scope for improvement and a time-bound ergonomic intervention is required, for Class-IV or "poor worksystem" refers to work condition not acceptable needing immediate ergonomic intervention, and for Class-V or "very poor worksystem" means a work condition is rejected and large-scale investment with intensive management involvement is required for improving the ergonomic performance.

Table 1.1 EPI grading of worksystems

Type of worksystems	Range of NTR	Grade	Type of worksystems
Class-I	85–100	Excellent	Maintain the prevailing work conditions
Class-II	70–84	Very good	Comparatively acceptable work condition; may initiate remedial steps wherever required
Class-III	50–69	Good	Acceptable work condition with a great scope for improvement; a time-bound ergonomic intervention required
Class-IV	45–49	Poor	Work condition not acceptable; needs immediate ergonomic intervention
Class-V	<45	Very poor	Work condition is rejected; large scale investment and/or intensive management involvement required

1.5 A Few Applications of EPI Model

The EPI model was verified and validated in several workplaces, such as blast furnace raw material division, merchant mill department for long products, LD2 department for flat products, equipment maintenance department, and power house#4 of the steel plant, and also in healthcare settings (Ray and Saha 2016). For illustration purpose, the method of computing EPI score, for two specific types of worksystems, viz., Equipment Maintenance Department (Support Services) and Merchant Mill Department (Long Products) of Tata Steel is explained below.

Application-I

For Equipment Maintenance Department (Support Services), the following factors are considered relevant for the purpose as described in Table 1.2.

The other factors as listed in Part-I are not found to be relevant in this section.

Computation of Normalized Total Rating (NTR)

With the ratings as assigned, the Normalized Total Rating of the emergency section is given by:

$$[\mathrm{NTR}] = \frac{\left[\sum_{i=1}^{n} \mathrm{SR}_i + m\right]}{n \times 18 + 10} \times 100,$$

where SR is the scale rating, i is a factor and n is the total number of factors, and m is the safety awareness (CO-SA) rating (0–10).

Let safety awareness (CO-SA) rating (0–10) = m = 8.

$$\text{Hence, Grand TS} = \left[\sum_{i=1}^{n} \mathrm{SR}_i + m\right] = 87 + 8 = 95$$

Maximum Scale Rating = $n \times 18 = 10 \times 18 = 180$.

Normalized Total Rating (NTR) in 0–100 scale is given by:

$[\mathrm{NTR}] = \frac{\left[\sum_{i=1}^{n} \mathrm{SR}_i + m\right]}{n \times 18 + 10} \times 100 = \frac{95}{180 + 10} \times 100 = 52.$

Referring to EPI Grading of worksystems in Table 1.1, it is found that a scale rating of 0.52 refers to a working condition that is “Good” from ergonomic perspective and represents an acceptable work condition with a great scope for improvement and a time-bound ergonomic intervention is required.

Application-II

For Merchant Mill Department (Long Products), the following factors are considered relevant for the purpose as described in Table 1.3.

Table 1.2 EPI factor influenced for equipment maintenance department (support services)

Item	Data collected or information available	EPI factor influenced	Rating
HC-3	Lifting of oil drum, lifting of ladder	F5. Frequency of lifting of weights	12
HC-4	Drawing trolley with motor/weight, pulling of brakes out of service, raising eels rapper motors to esp top, pulling of chain block chain with weight	F6. Force required to push or pull objects	9
HC-5	Dragging of trolley with materials, putting breakers in service position, some valve operation	F7. Movements of human body	6
PE-2	Turbine control at BCD, electrical control at ECR, any job in basement, equipment overhauling	F8. Assessment of visual environment in the workplace	6
PW-2	Esp top gear box, inspection of motors, Tu hall operator, thermal control, BCD operation, electrical operation at electrical control, panel overhauling, boiler control operation, thermal control room operation unit, old thermal room operationTrf overhauling, pump house operation, use of ladder, basement floor movement, etc.	F9. Engineering anthropometry	9
PW-2	ESP emulator changing, changing of pumps and motors, working in shift area	F10. Work posture	9
PE-4	Loco operator's cabin, loco operation in Ld & Bl furnace area	F11. Assessment of thermal environment in the workplace	15
HC-8	Cutting with hacksaw, grinding, use of drill, control room operators-controlling boilers, controller in electrical control room, boiler controllers controlling boiler operation, men at electrical control room, controlling at ECR, boiler control	F14. Repetitive motions/frequent use of hand tools/both hands and feet operating/same posture/information overload/insufficient time to sense and respond to signals/physical fitness/knowledge of training	12
PE-1	Load testing of diesel engine	F15. Assessment of auditory environment	9
Total			87

Table 1.3 EPI factor influenced for merchant mill department (long products)

Item	Data collected or information available	EPI factor influenced	Rating assigned
HC-6	Packing and strapping of furnished base, scale removal from the furnace, removal of cutting from cold shear	F1. Pace or speed of work under the control of the operator	12
HC-7	Pushing out the billets from the furnace, crane drives in finishing and stripping section, pulpit B operates while seeing loop in the bar	F2. Adequacy of fatigue allowances for jobs	6
HC-3	Removal of bullet front and cuttings, removal of scale from mill tunnel, sorting on the hold up bed	F5. Frequency of lifting of weights	6
HC-4	Billet tilting on conveyer roll prior to furnace, putting motor house breaker in or taking out, pulling and pushing oxygen and weight cylinder trolley	F6. Force required to push or pull objects	6
HC-5	Seeing the entry and exit guide setting, alignment of TMT heeding, alignment of pendulum shear with roughing	F7. Movements of human body	6
PE-2	Face area, gas lines water seal, etc., cold shear area, bearing shop for roll assembly, billet yard	F8. Assessment of visual environment in the workplace	9
PW-1	Cleaning and maintenance below gearbox, cleaning below cold sheets and cold sheet 2	F10. Work posture	6
PE-4	Charging bed conveyer roll area prior to furnace, billet pushing cabin, billet withdrawing cabin, crane operator cabin, pushing cabin on the backside of the furnace, crossover bridge for inspection of cooling bed, crane drivers cabin, withdrawal and twist pinch roll operator's cabin, cold sheet operator rear sheet	F11. Assessment of thermal environment in the workplace	9
PE-5	Grinding operation while guides are grinded, face area hydraulic purees platform, TMT heads presume gauge display	F12. Workers complain about physical environment in their workplaces	6

(continued)

Table 1.3 (continued)

Item	Data collected or information available	EPI factor influenced	Rating assigned
HC-8	Cutting from behind near cold sheet bars to be removed and sorted manually, twist pinch roll operator, twist pinch roll operator critically operates control using both hands, changing bed pulpit operator, motor house substation in charge, pulpit operator. This person when asked to tilt the billet which was struck midway inside the furnace, when new system in hydraulic/pneumatic was commissioned, all the people did not have appropriate knowledge of training	F14. Repetitive motions/frequent use of hand tools/both hands and feet operating/same posture/information overload/insufficient time to sense and respond to signals/physical fitness/knowledge of training	9
PE-1	Face area, cold sheet 1, cold sheet 2, roller's cabin where grinding of guides is done, big diameter base moving on the roller from the shuffle bar	F15. Assessment of auditory environment	6
Total			81

The other factors as listed in Part-I are not found to be relevant in this section.

Computation of Normalized Total Rating (NTR)

With the ratings as assigned, the Normalized Total Rating of the emergency section is given by:

$$[\mathrm{NTR}] = \frac{\left[\sum_{i=1}^{n} \mathrm{SR}_i + m\right]}{n \times 18 + 10} \times 100,$$

where *SR* is the scale rating, *i* is a factor and *n* is the total number of factors, and *m* is the safety awareness (CO-SA) rating (0–10).

Let safety awareness (CO-SA) rating (0–10) = m = 5.

$$\text{Hence, Grand TS} = \left[\sum_{i=1}^{n} \mathrm{SR}_i + m\right] = 81 + 5 = 86$$

Maximum Scale Rating = $n \times 18 = 11 \times 18 = 198$.

Normalized Total Rating (NTR) in 0–100 scale is given by: $[\mathrm{NTR}] = \frac{\left[\sum_{i=1}^{n} \mathrm{SR}_i + m\right]}{n \times 18 + 10} \times 100 = \frac{86}{198 + 10} \times 100 = 45.$

Referring to EPI Grading of worksystems in Table 1.1, it is found that a scale rating of 0.452 refers to a working condition that is "Poor" from ergonomic perspective and represents an unacceptable work condition which needs immediate ergonomic intervention.

With the application of generic EPI model of several worksystems of Tata Steel, a number of observations regarding the usefulness of the model that may be made are (i) the EPI model is applicable to all types of worksystems, (ii) with the use of EPI model on a continuous basis, there is a high probability that the persons at all levels of organization become aware of importance of ergonomic design for sustainable worksystem performance, (iii) as the model is factor specific, ability of the persons concerned to judge the ergonomic performance from several "man–machine" perspectives is expected to improve in near future, (iv) the EPI model is cost-effective in the sense that it mainly suggests preventive measures for improving ergonomic performance and with the implementation of preventive measures, the effect of occupational risk factors get minimized.

1.6 Concluding Remarks

The EPI model has been found to be very effective and a team of qualified ergonomics and human factors professionals from ergonomics and safety department of Tata Steel is responsible for computing EPI score of a given worksystem as and when required. The role of such professionals lies in assessing the present level of ergonomic performance and identifying the deficient areas (the list of factors where the scores are low) and the deficient areas where ergonomic interventions are desired. It has been observed that such ergonomically deficient areas are regularly identified and in terms of ergonomic intervention projects are being undertaken by these professionals. In certain areas, external expertise in the form of hiring consultants is needed for undertaking such improvement initiatives.

References

Ray PK, Saha E (2016) Ergonomic performance measurement system of work-systems in healthcare. In: Vincent GD, Nancy L (eds) Advances in human factors and ergonomics in healthcare, Proceedings of the Applied Human Factors and Ergonomics (AHFE) 2016 International Conference on Human Factors and Ergonomics in Healthcare, Springer, pp 329–342

Ray PK, Tewari VK (2012) Design and implementation of ergonomic performance measurement system at a steel plant in India, work. J Prev Assess Rehabil 41:5943–5949

Chapter 2
Decision Tree for Manual Material Handling Tasks Using WEKA

R. Rajesh, J. Maiti and M. Reena

Abstract Manual material handling (MMH) is one of the most physically demanding operations where workers are exposed to repetitive movements, awkward postures, contact stresses, and forceful exertions. MMH results in biomechanical and physiological strain on material handlers. Numerous observational and direct methods are used for assessing MMH tasks. In industrial settings, observational methods are best suited for ergonomic assessments of MMH tasks. But issues such as unclear classification levels, need of expertise, obstructive and invasive nature of data collection procedures, and time and cost requirements create difficulty for safety and ergonomic engineers in using observational methods. The objective of the study is to propose a decision aid to help safety and ergonomic engineers in making an easy ergonomic assessment of MMH tasks for the case study plant in West Bengal. In the current study, WEKA is used to classify the MMH tasks using J48 algorithm. The input data for this study is obtained from our previous work on the development of Cube model-2 (Rajesh et al. in IIE Trans Occup Ergon Human Factors, 2(1):39–51, 2014), in which a field survey of MMH tasks is conducted and classified the MMH tasks into three categories. The output MMH task classifications in WEKA are also classified into three levels. Overall true-positive rate of 0.813, false-positive rate of 0.170, and ROC value of 0.851 are obtained. The weighted Kappa statistic is 0.715. The results from WEKA are encouraging and enable us to use the simple decision tree to judge the physical demands of material handlers. The practical relevance of the study is that the decision tree is helpful for industry practitioners in assessing the MMH tasks therein.

Keywords Manual material handling · Cube model-2 · Data mining · WEKA · Decision tree

---

R. Rajesh (✉) · M. Reena
Rajiv Gandhi Institute of Technology, Kottayam, Kerala, India
e-mail: rrajeshraghu@gmail.com

J. Maiti
Indian Institute of Technology, Kharagpur, West Bengal, India

P.K. Ray and J. Maiti (eds.), *Ergonomic Design of Products and Worksystems—21st Century Perspectives of Asia*, Managing the Asian Century,
https://doi.org/10.1007/978-981-10-5457-0_2

2.1 Introduction

Manual material handling (MMH) is an integral part of many industrial activities and is a major contributor of musculoskeletal disorders. MMH activities involve high physical demands, awkward postures, contact stresses, and forceful exertions, the contents of which vary significantly depending on the work situations and industrial sectors. MMH jobs previously composed of a high percentage of lift–lower–carry operations have gradually been replaced by pushing–pulling operations. The evolving nature of MMH exposure calls for not only better exposure measurement and assessment strategies, but also for methods that are easy use and implement.

Worksystem factors relevant for MMH exposure include worker characteristics, material characteristics, Task/workplace characteristics, and environment/organizational characteristics (Rajesh et al. 2014; Rajesh 2016). The physiological and biomechanical responses need to be examined to quantify the physical load and strain arising from the MMH exposure. For undertaking interventions to address ergonomic issues in workplaces engineers and Ergonomists resort to observational and direct measurement tools to quantify these responses (Takala et al. 2010). Ergonomic observational tools for assessing combined MMH tasks that are suitable for the on-field industrial environment are scarce, and newer tools are needed to be explored toward making ergonomic evaluations easier and simpler. For safety and health practitioners the current need is to make quick and easy qualitative ergonomic assessment of the work activity. These must complement the observational and quantitative ergonomic methods that are in place in the industries.

The objective of this paper is to highlight a decision tree devised for a case study plant. The methods section of the paper gives a brief description of the case study plant and description of ergonomic evaluation done by existing ergonomic tools. In the next section, the performance of decision tree with respect to the ergonomic tools is presented. Finally, conclusion section presents the performance summary, its limitation, and scope for future work.

2.2 Method

2.2.1 Dataset

The study was conducted in a bearing manufacturing plant in India. Ball bearing and roller bearings are the main products produced by the manufacturing plant. The plant has a diverse material handling system catering to the specific needs of different sections. The material handling system configuration coupled with worksite (i.e., location) layout, task–workplace characteristics shall produce different physical exposure doses on the material handlers. The work sampling study was conducted over 19 locations in Store section of the plant. At the completion of the study, 143 MMH tasks and 262 task elements were observed. Each of these MMH tasks was composed of multiple MMH task elements, consisting of lifting–lowering, carrying, and pushing–pulling activities. For each of the MMH tasks observed from the observational study, a biomechanical analysis is undertaken. Biomechanical analysis was done using digital human models in CATIA software (Version 5, Dassault Systems, and France). The most frequent posture observed for a particular activity was used for the analysis. Biomechanical analysis of postural activities was carried out for the starting and ending postures for each task element, and was limited to lower back and shoulder joints. To account for the time dimension, the cumulative loading concept (Callaghan et al. 2001) was applied over the representative static postures of the task elements. Details of the biomechanical analysis in Cube Model-2 can be found in Rajesh et al. (2014). Based on the instantaneous loads and the cumulative dose the exertion level for each task is classified into 'low', 'medium', or 'high' as per the biomechanical criteria (Rajesh et al. 2014). The exertion level 'low' implies that biomechanical load does not cause physical strain on the material handler and is acceptable. The exertion level 'medium' implies that the biomechanical load could cause physical strain when done continuously. Therefore, the task needs to be accepted conditionally. The exertion level 'high' implies that there is a high biomechanical load that would cause physical strain to the material handler, and hence is unacceptable. The data set so obtained from the biomechanical analysis forms the basis of this study. Details of the dataset are given in Table 2.1.

2.2.2 *Decision Tree*

A decision tree is a decision-modeling tool that graphically displays the classification process of a given input for given output class labels. The decision tree mechanism is transparent and allows one to follow the tree structure easily to see how the decision is made. A decision tree is a tree structure consisting of internal and external nodes connected by branches. An internal node is a decision-making unit that evaluates a decision function to determine which child node to visit next. The external node, on the other hand, has no child nodes and is associated with a label or value that characterizes the given data. A decision tree is a top-down induction method of classification that has three main steps. First, the root node at the top node of the tree considers all samples and passes through the samples information in the second node called 'branch node'. The branch node generates rules for a group of samples by considering all attribute values and finalizes the decision rule by pruning. After fixing the best rule, the branch nodes send the final target value in the last node called the 'leaf node'. Finally, it chooses the attribute that offers rules with minimum error and constructs the final decision tree. Decision tree classifier techniques have been used successfully for a wide range of classification problems, but none so far in MMH application. Software used in his study is

Table 2.1 Details about dataset

Attributes	Description	Type (unit)	Descriptive
Load	Total material moved manually between locations	Numeric (kg)	Min 5 to max 2625 kg. Mean 405 & Std Dev 454
Unit weight	Unit material handled manually	Numeric (kg)	Min 1 to max 1044 kg. Mean 155 & Std Dev 235
MMH	Type of MMH task, i.e., lifting or lowering, carrying, pushing or pulling	Nominal (no)	Lift–lower (n = 125), Carry (36), Push–pull (101)
Worker no	Number of workers undertaking the MMH task	Numeric (no)	Min 1 to max 4. Mean 1.447 & Std Dev 0.669
Working_height	Working height at which material is handled	Nominal (no)	Floor to shoulder height (n = 6), shoulder height (22), floor height (18), floor to waist height (97), waist to shoulder height (69), waist height (50)
Distance	Horizontal distance moved during MMH task	Numeric (m)	Min 0 to max 200 m. Mean 20.782 & Std Dev 40.824
Repetition	Number of times MMH activities are repeated	Numeric (no)	Min 1 to max 480. Mean 21.836 & Std Dev 44.818
Duration	Duration of the continuous MMH activity	Numeric (minutes)	Min 1 to max 32 min. Mean 7.378 & Std Dev 7.406
Exertion	Workers exertion level due to MMH task	Nominal (no)	Low (n = 72), Medium(147), High(43)

Waikato Environment for Knowledge Analysis (WEKA). WEKA has implemented data mining algorithms using the JAVA language. It is open-source software and contains tools for data preprocessing, classification, regression, clustering, association rules, and visualization.

2.2.3 *Performance Measures*

There are several terms that are used for estimating the performance measure. They are true-positive (TP), true-negative (TN), false-negative (FN), and false-positive (FP). The performance of the predictive model is calculated based on the precision, recall values of classification matrix. Precision is the fraction of retrieved instances that are relevant. Recall is a fraction of relevant instances that are retrieved.

True-positive rate:

$$\mathrm{TPR} = \frac{\mathrm{TP}}{(\mathrm{TP} + \mathrm{FN})}. \tag{2.1}$$

True-negative rate:

$$\mathrm{FPR} = \frac{\mathrm{FP}}{(\mathrm{FP} + \mathrm{TN})} \tag{2.2}$$

$$\mathrm{Precision} = \frac{\mathrm{TP}}{(\mathrm{TP} + \mathrm{FP})} \tag{2.3}$$

$$\mathrm{Recall} = \frac{\mathrm{TP}}{(\mathrm{TP} + \mathrm{FN})}. \tag{2.4}$$

Other performance measures obtained from WEKA include Receiver operating characteristic (ROC) and Kappa statistic and root-mean-square error (RMSE). ROC provides comparison between predicted and actual target values in a classification. It describes the performance of a model with a complete range of classification thresholds. ROC area varies between 0 and 1 intervals. An increasing value indicates better classification, with an area of one representing perfect classification. The Kappa Statistic can be defined as measuring the degree of agreement between two sets of categorized data. Kappa statistic varies between 0 and 1 intervals; the higher the value, the stronger the agreement (Sim and Wright 2005). Root-mean-square error is defined as the square root of the sum of squares error divided by a number of predictions. It measures the differences between values predicted by a model and the values actually observed. The smaller the RMSE, the better the accuracy of the model. The performance of the learning techniques is dependent on the nature of the training data.

The current study is classified into three classes. For the multiclass problem, sok09 has presented a number of performance measures. The classifiers are evaluated by some the multiclass performance measures listed below.

Average accuracy:

$$\mathrm{ACC} = \frac{\sum_{i=1}^{\mathrm{N}} \frac{\mathrm{TP}_i + \mathrm{TN}_i}{\mathrm{TP}_i + \mathrm{FN}_i + \mathrm{FP}_i + \mathrm{TN}_i}}{\mathrm{N}}. \tag{2.5}$$

Average classification error:

$$\mathrm{ER} = \frac{\sum_{i=1}^{\mathrm{N}} \frac{\mathrm{FP}_i + \mathrm{FN}_i}{\mathrm{TP}_i + \mathrm{FN}_i + \mathrm{FP}_i + \mathrm{TN}_i}}{\mathrm{N}}. \tag{2.6}$$

An average per-class agreement of the data class labels with those of classifiers:

$$\text{Precision}_{\text{M}} = \frac{\sum_{i=1}^{\text{N}} \frac{\text{TP}_i}{\text{TP}_i + \text{FP}_i}}{\text{N}}. \tag{2.7}$$

An average per-class effectiveness of a classifier to identify class labels:

$$\text{Recall}_{\text{M}} = \frac{\sum_{i=1}^{\text{N}} \frac{\text{TP}_i}{\text{TP}_i + \text{FN}_i}}{\text{N}}. \tag{2.8}$$

Relation between data positive labels and those given by a classifier based on a per-class average:

$$\text{Fscore}_{\text{M}} = \frac{(\beta^2 + 1)\text{Precision}_{\text{M}}\text{Recall}_{\text{M}}}{\beta^2\text{Precision}_{\text{M}} + \text{Recall}_{\text{M}}}. \tag{2.9}$$

Matthews correlation coefficient:

$$\text{MCC} = \frac{\text{F}^2 + (\text{N} - 2)\text{TF} - (\text{N} - 1)\text{F}^2}{(\text{T} + (\text{N} - 1)\text{F})^2}. \tag{2.10}$$

Weighted Kappa statistic is obtained using the weights assigned for each class (Eq. 2.11). In the case of a task with true exertion 'Low' and classified as 'Medium' there is no physical strain on the worker, i.e., ergonomic cost is nil. It is likely that an operating cost exists if the ergonomist undertakes interventions, and on the other hand, a task with true exertion 'High' and classified as 'Medium'; there is a physical strain on the worker, i.e., ergonomic cost is present but ignored by the classifier. The weights are assigned based on author's judgment regarding penalties for incorrect classification and are shown in Table 2.2.

Weighted Kappa statistic based on ergonomic weights is as follows:

$$\text{K}_{\text{w}} = \frac{\text{P}_{\text{e(o)}} - \text{P}_{\text{e(w)}}}{1 - \text{P}_{\text{e(w)}}}, \tag{2.11}$$

Table 2.2 Weights for Kappa agreement analysis

Exertion level by biomechanical analysis	Exertion level predicted by classifier		
	Low	Medium	High
Low	1	0.75	0
Medium	0.25	1	0.75
High	0	0.25	1

where observed weighted proportion of agreement is

$$P_{e(o)} = \sum_{i=1}^{N=3} \sum_{j=1}^{N=3} w_{i,j}\, p_{i.} p_{j}.$$

Chance of expected weighted proportion of agreement is

$$P_{e(w)} = \sum_{i=1}^{N=3} \sum_{j=1}^{N=3} w_{i,j}\, p_{i.} p_{j}.$$

2.2.4 WEKA Procedure

In WEKA, all data is considered as instances and features in the data are known as attributes (Witten et al. 1999; Hall et al. 2009). The dataset with nine attributes (see Table 2.1) is preprocessed in WEKA and stored in a database. The data file normally used by WEKA is in ARFF file format, which consists of special tags to indicate different things in the data file. The refined data is given to the proposed prediction models as input, predicted class will be extracted and performance evaluation metrics of different algorithms. In this study J48, Random forest, REP and LMT algorithm are employed. For the test option tenfold cross-validation is used. Here the dataset is split by tenfold, each fold contains 90% of the samples to construct a model and the remaining 10% is used to evaluate the model performance. The performance measures from WEKA include TPR, FPR, Correct classification (%), MCC, ROC area, Kappa statistic, and RMSE. Confusion matrices are very useful for evaluating classifiers. In the confusion matrix, the columns represent the predictions, and the rows represent the actual class. Based on it the multiclass performance measures ACC, ER, $Precision_M$, $Recall_M$, F_{score}, MCC_M, and K_w are estimated (Sokolova and Lapalme 2009; Jurman et al. 2012).

2.3 Results and Discussion

2.3.1 Decision Tree Model Results

The results from J48, Random forest, and REP and LMT algorithm are shown in Table 2.3. We observe using the TPR and TNR measures that J48 was the best performing algorithm and the second best was Random Forest. By examining correct classification measure J48 performed best. Kappa statistic indicates that J48 and the Random Forest have a substantial agreement between biomechanical analysis and the classifiers. J48 returned the highest correct classification (81%) and

Table 2.3 Prediction performance measures (n = 262)

Performance measures	Algorithm			
	J48	Random forest	REP	REP
TPR	0.813	0.794	0.634	0.76
FPR	0.17	0.19	0.347	0.213
Precision	0.814	0.786	0.603	0.759
Recall	0.813	0.794	0.634	0.757
Correct classification (%)	81.3	79.4	63.3	75.9
ROC area	0.851	0.884	0.725	0.846
Kappa statistic	0.668	0.632	0.287	0.573
RMSE	0.332	0.324	0.418	0.361

Table 2.4 Confusion matrix from J48 and random forest algorithm in WEKA

Exertion level by biomechanical analysis	Predicted exertion level by J48			Predicted exertion level by random forest		
	Low	Medium	High	Low	Medium	High
Low	61	11	0	62	10	0
Medium	2	130	15	7	128	12
High	0	21	22	0	25	18

Table 2.5 Prediction performance using multiclass measures (n = 262)

Performance measures	Algorithm			
	J48	Random forest	REP	LMT
Accuracy (ACC)	0.872	0.856	0.823	0.829
Error rate (ER)	0.13	0.144	0.177	0.17
Precision_M	0.788	0.761	0.734	0.73
Recall_M	0.748	0.717	0.746	0.693
Fscore	0.768	0.738	0.74	0.711
Matthews correlation coefficient (MCCM)	0.563	0.536	0.446	0.473
Weighted Kappa (K_w)	0.715	0.649	0.647	0.631

the lowest RMSE. ROC values for J48, Random Forest, and LMT algorithm indicated good accuracy (ROC > 0.8). It appears that REP and LMT classifiers are not suitable for classifying MMH tasks as their correct classification %, Kappa statistic, and RMSE are lower than J48 and Random Forest algorithms. Confusion matrix from J48 and Random Forest algorithm is presented in Table 2.4. Though WEKA performance measures for J48 and Random Forest indicate only slight advantage for J48, multiclass measures in Table 2.5 indicate a distinct advantage for J48 classifier. Weighted kappa statistic for J48 is 0.715 against 0.649 for Random Forest, both of which indicates substantial agreement (Sim and Wright 2005). The lower value for Random Forest is due to the higher number of misclassification by Random forest as compared to J48 (Table 2.4). Among the four classifiers REP performed the worst and J48 returned the best results. Hence, J48 is more suited for prediction purpose.

2.3.2 Decision Tree Induced MMH Guidelines

Figure 2.1 shows the decision tree obtained using J48 algorithm. Figure 2.1 shows a simple decision tree that consists of a decision node, branches, and leaves. The root node in this example is 'MMH'. The results of the test at the root node (whether the MMH task type is 'push' or others) cause the tree to split into branches. The two branches in this case are '=push' indicating 'push–pull' MMH task and '! = push'.

Indicate 'lift-lower-carry' MMH task. The leaves are the decisions made and are found on at the end of the last branch. The size of the tree is 37 with 19 leaves. For example, the decision made for the task attributes, MMH = push (the task is push–pull MMH task), Repetition ≤ 1 (one time task), Duration = 4 min with Load ≤ 462 kg (material load using hand pallet truck is less than 462 kg), is 'Low'. Some of the useful decision rules that can be practically used in the case study company are presented in terms of MMH guidelines.

- Pushing tasks that are repeated more than 3 times are categorized as high exertion task and are unacceptable. An immediate solution to the above condition is to take a break after every 3 continuous push–pull task.
- Lifting–carrying–lowering tasks that are repeated more than 200 times are categorized as high exertion task and are unacceptable. An immediate solution to the above condition is to reduce repetitions below 200 or take a break at the end of 200 repetitions.
- In the case of lifting–carrying–lowering tasks where unit weight is above 36 kg and total material handling load is more than 80 kg, the MMH task is categorized as high exertion task and is unacceptable. An immediate solution to the above condition is to reduce the total MMH load to below 80 kg. In addition, a reduction of unit weight from 36 kg is likely to reduce the stress further and make the MMH task a medium exertion task. This can be seen from the node 'Unit Weight $\leq$ 36' to 'Load ≤ 300'.
- In the case of lifting–carrying–lowering tasks where unit weight is above 24 kg, total material handling load is more than 682 kg and working height is floor–waist height, the MMH task is categorized as high exertion task and is unacceptable. An immediate solution to the above condition is to reduce total MMH load to below 682 kg. This can be seen from the node 'Unit Weight > 24' to 'Load $\leq$ 682'. In addition, a reduction of unit weight to below 24 kg is likely to reduce the stress further and make the MMH task a medium exertion task.

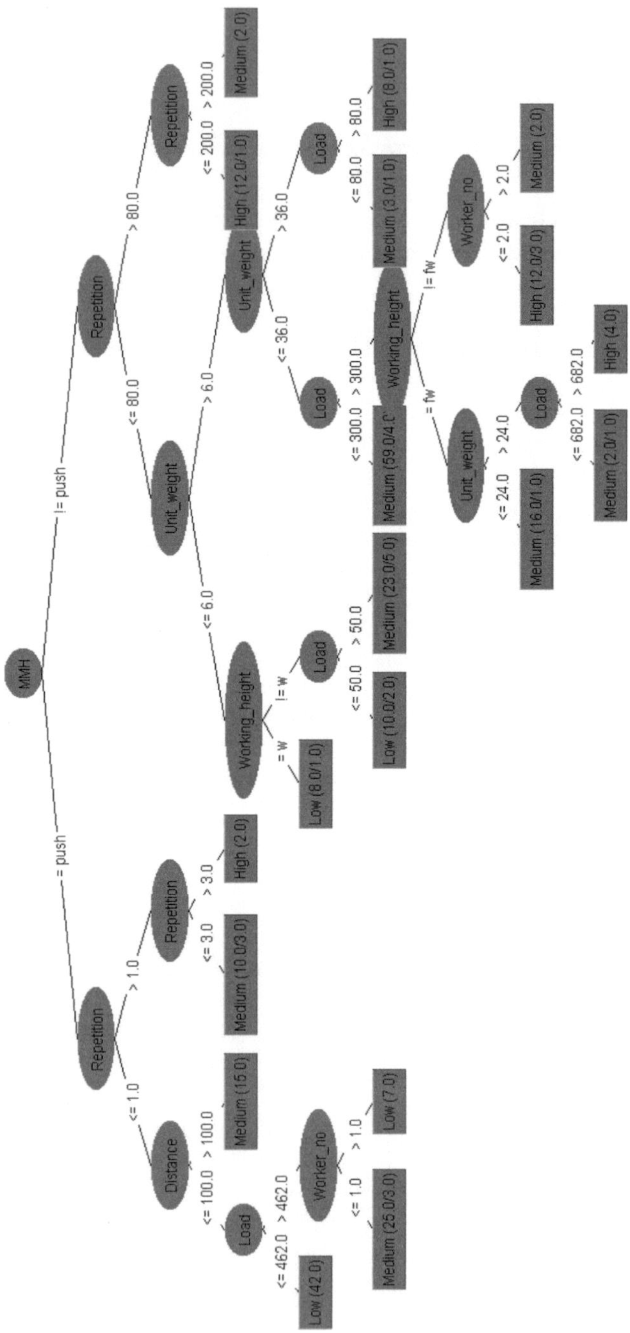

Fig. 2.1 Decision tree using J48 algorithm in WEKA

- In the case of lifting–carrying–lowering tasks where unit weight is between 25 and 36 kg, total material handling load is more than 300 kg and with 2 or lesser worker group; the MMH task is categorized as high exertion task and is unacceptable. An immediate solution to the above condition is to increase the number of workers in the group.

2.4 Conclusion

The output MMH task classifications in WEKA are also classified into three levels using WEKA-based classifiers J48, Random Forest, REP, and LMT. J48 and Random Forest performed better than the other two. Multiclass performance measures indicated J48 to be better than Random Forest classifier. Weighted Kappa statistic was the highest for J48. Among the four classifiers REP performed the worst and J48 gave the best results. J48 returned 81% correct classification. Overall true-positive rate of 0.813, false-positive rate of 0.170, and ROC value of 0.851 are obtained. The weighted Kappa statistic is 0.715. The results from WEKA are encouraging and enable us to use the simple decision tree to judge the physical demands of material handlers. The decision tree using J48 not only helps the industry practitioner in making quick ergonomic assessment of MMH task occurring in the plant but also provides insights toward intervention measured toward reducing the physical load in 'high' exertion tasks.

The practical relevance of the study is that the decision tree is helpful for industry practitioners in assessing the MMH tasks in the case study plant. The limitation of the study is that the sample size is small. Reasons for the misclassification need to be examined.

Acknowledgements The authors also appreciate the support provided by Mr. Sukanta Chanda, Dr. O.B. Krishna and all the workplace personnel involved during the field study.

References

Callaghan JP, Salewytsch AJ, Andrews DM (2001) An evaluation of predictive methods for estimating cumulative spinal loading. Ergonomics 44(9):825–837

Hall M, Frank E, Holmes G, Pfahringer B, Reutemann P, Witten IH (2009) The WEKA data mining software: an update. ACM SIGKDD Explorations Newsletter 11(1):10–18

Jurman G, Riccadonna S, Furlanello C (2012) A comparison of MCC and CEN error measures in multi-class prediction. PloS One7(8):e41882

Rajesh R (2016) Manual material handling: a classification scheme. Procedia Technol 24:568–575

Rajesh R, Maiti J, Samanta B (2014) Application of the cube model for biomechanical exposure assessment of combined manual material handling tasks in a manufacturing plant in India. IIE Trans Occup Ergon Human Factors 2(1):39–51

Sim J, Wright CC (2005) The kappa statistic in reliability studies: use, interpretation, and sample size requirements. Phys Ther 85(3):257–268

Sokolova M, Lapalme G (2009) A systematic analysis of performance measures for classification tasks. Inf Process Manage 45(4):427–437

Takala E, Pehkonen I, Forsman M, Hansson G, Mathiassen SE, Neumann WP, Sjogaard G, Veiersted KB, Westgaard RH, Winkel J (2010) Systematic evaluation of observational methods assessing biomechanical exposures at work. Scand J Work Environ Health 36(1):3–24

Witten IH, Frank E, Trigg LE, Hall MA, Holmes G, Cunningham SJ (1999) Weka: practical machine learning tools and techniques with Java implementations. Working Paper 99/11, University of Waikato, New Zealand

Chapter 3
Interventions for Combined Manual Material Handling Tasks Based on Biomechanical and Physiological Responses

R. Rajesh and J. Maiti

Abstract Manual material handling (MMH) exposure arises from the work system characteristics leading to strain on musculoskeletal, cardiovascular and neurological system. A combined manual material handling task is composed of a combination of tasks involving lifting, lowering, pushing, pulling or carrying. The paper highlights interventions for a MMH work system in a bearing manufacturing plant in India through an experimental study where workers are faced with both biomechanical and physiological strains while undertaking combined MMH tasks. The intervention procedure proposed for combined manual material handling tasks in this study accounts for both biomechanical and physiological strains. The biomechanical measures considered for interventions are 'Body Part Discomfort Frequency Severity' and 'Muscle effort', and physiological measures considered for interventions are 'energy expenditure rate', 'heart rate' and 'rating of perceived exertion'. The design solutions have addressed the whole-body exertions faced by the material handlers. The practical implication of the study is that it has highlighted a few work system characteristic that produces less biomechanical and physiological stresses. Several recommendations for improved work system factors (e.g. load, equipment, and workplace) for combined loading, push–pull and unloading tasks are provided. The recommended interventions are generic and can be used in any MMH work system where human effort is dominant, and level of mechanization is low. The values obtained from this simulation are only first approximation and need to be verified through actual implementation of the interventions.

Keywords Combined manual material handling · Fractional factorial experiment · Biomechanics · Physiology · Interventions

---

R. Rajesh (✉)
Rajiv Gandhi Institute of Technology, Kottayam, Kerala, India
e-mail: rrajeshraghu@gmail.com

J. Maiti
Indian Institute of Technology, Kharagpur, West Bengal, India

P.K. Ray and J. Maiti (eds.), *Ergonomic Design of Products and Worksystems—21st Century Perspectives of Asia*, Managing the Asian Century,
https://doi.org/10.1007/978-981-10-5457-0_3

3.1 Introduction

A combined manual material handling task is composed of a combination of tasks involving lifting, lowering, pushing, pulling or carrying. Over the past few decades, MMH has changed due to the availability of material handling equipment (MHE) and increased use of mechanization. Work system conditions influence the physical demand during material handling activity. MMH exposure arises from the characteristics of the work system components, i.e. worker characteristics, material characteristics, task characteristics, environmental and organizational characteristics (Ciriello 2005, 2007; Kuijer et al. 2000; Rajesh and Maiti 2011; Rajesh et al. 2014; St-Vincent et al. 2005; Young et al. 2013). The ergonomic risk factors present in the system lead to strain on musculoskeletal, cardiovascular and neurological system. The complex interaction between ergonomic stress factors in a worksystem poses two challenges: the first one is the identification and quantification of MMH exposure, and the second one is in undertaking design or redesign of MMH tasks that can reduce the severity of manual material handling.

Interventions must account for biomechanical and physiological strains within an MMH worksystem. From a risk perspective, the MMH exposure produces varying effects on different body parts. Within each MMH task component interventions are useful only when significant effects can be ascertained, so that design options can be prioritized. Notwithstanding the advances in exposure modelling and intervention research (Chaffin 2005; Wells et al. 2007; Hu et al. 2011; Yun et al. 2013; Perez and Neumann 2015; Village et al. 2015), there is still a need to focus on methodological aspects of interventions accounting for ergonomic stressors. To model MMH exposure and redesign MMH task, a careful ergonomic evaluation through biomechanical, physiological and psychophysical measurements and analysis is required. Biomechanical stresses are rather instantaneous and arise due to high force and postural demands. Cardiovascular stress arises from MMH exposures over the longer time dimension. Within a worksystem where combinations of MMH task types are undertaken, there is a need to examine (i) factor effects and MMH exposure pathway, and (ii) effects on body parts, within the intervention method. Therefore, a careful ergonomic evaluation that is linked with the intervention method would then help in identifying and prioritizing design solutions for worksystems with multiple risk factors.

This paper focuses on highlighting few interventions for an MMH worksystem in a bearing manufacturing plant in India where workers are faced with both biomechanical and physiological strains (Rajesh 2015). Rajesh et al. (2014) in their field study found that MMH exposure leads to significant biomechanical exposure. Some of those task durations were observed to be more than 30 min. The continuing experimental study did highlight the biomechanical and physiological demands of the combined MMH tasks (Rajesh 2015). Further, they found that MMH exposure leads to strain on different body parts (Rajesh 2015). The primary focus of their work (Rajesh et al. 2014; Rajesh 2015) has been to extend in two areas: (i) exposure quantification accounting for 'time' factor, i.e. cumulative biomechanical and physiological demands and (ii) development of indicators for whole-body strain. The intervention issues focused upon in this paper include (i) how to undertake interventions when both biomechanical and physiological demands occur in an MMH system, (ii) which of the task factors needs to be focussed upon and (iii) how to prioritize interventions based on risks on body parts.

3.2 Method

3.2.1 Details of Experimental Study

A fractional factorial experiment with ten factors each at two levels was conducted in a bearing manufacturing plant. The independent variables were 'Weight', 'Distance', 'Pace', 'Cycle', 'Height', 'Team', 'Combined', 'Direction', 'Motion pattern' and 'MHE'. The dependent variables considered were 'muscle effort', 'change in heart rate', 'discomfort rating' and 'ratings of perceived exertion'. Four male subjects were randomly selected from the general population to take part in the study. Informed consent for the procedure, questionnaire and data collections sheet by the management of the organization were obtained prior to the actual data collection. The experimental trials (22 × 4) were randomized. These are referred as worksystem settings T1 to T22. The task performed by the material handler consisted of lifting–lowering–carrying 30 boxes from storage rack onto material handling equipment, then by pushing–pulling the material handling equipment, the load was moved by a fixed distance, and finally by lifting–lowering–carrying, the load was unloaded into storage storage rack. The MMH task performed by the material handler was video recorded by six cameras into a digital video recorder at 30 Hz, and the corresponding real-time hand forces from LUTRON force gauge were recorded into a laptop. Heart rate was measured using Polar heart rate monitor at intervals of 15 s. HR data was downloaded onto a personal computer at the end of the experiment for further analysis. On completion of the task, the psychophysical responses of discomfort and exertion were measured using body part discomfort scale (Olendorf and Drury 2001) and CR10 scale (Borg 2004). For each trial, 'muscle effort' (Ayoub 1998; Rajesh 2015) was estimated using a novel methodology devised for measuring biomechanical exposure from combined MMH tasks, and energy expenditure was estimated using Garg's model (Garg et al. 1978). Details of the muscle effort methodology can be found in Rajesh (2015).

Muscle effort model estimates the whole-body effect for undertaking a task.

Muscle effort:

$$\mathrm{ME} = \int_{t=0}^{T} \sum_{i=1}^{7} \left(\frac{\mathrm{M_i(t)}}{\mathrm{S_i(t)}} \right)^2 \mathrm{dt}, \tag{3.1}$$

where S_i is the moment strength of each joint, M_i is the reactive moment at each joint, n is 7 for whole body (wrist, elbow, shoulder, back, hip, knee, knuckle), T is the duration of the task.

The physiological response variable namely change in heart rate (ΔHR) is the difference in mean heart rates recorded for each component of the MMH cycle:

$$\Delta HR_{i,j} = HR_0 - HR_{i,j}, \tag{3.2}$$

where HR_0 is the rest heart rate, HR_i is the mean of recorded heart rate for ith component of the jth cycle in an MMH trial.

3.2.2 Intervention Procedure

The overall intervention strategy is summarized as follows:

- The basis of intervention (biomechanical) is based on discomfort scores recorded. The severity of discomfort is evaluated on a six-point ordinal scale starting from 0 to 5, and the corresponding verbal categories are 'no discomfort', 'light or perceivable discomfort', 'low discomfort', 'moderate discomfort', 'high or more than moderate discomfort' and 'extreme or very uncomfortable ''. The discomfort for each body segment is given by its discomfort score (BPDS). The overall whole-body discomfort score for each of the trials is given by Body Part Discomfort Frequency Severity (BPDFS). To identify the body segments that are stressed BPDS is used, and to identify overall (cumulative) whole-body biomechanical exposure effect BPDFS is used. In deciding the intervention, the criteria of Body part discomfort score (BPDS) >2 are used. The focus of intervention needed is decided by examining BPDFS. A higher BPDFS indicates the greater stress on the body segment and needs priority. Interventions are prioritized based on the BPDFS ratings for upper extremity (UE), lower back (LB) or lower extremity (LE).
- The basis of intervention (physiological) is based on the change in heart rate and energy expenditure. The overall physiological effect of variables is obtained by examining the change in heart rate (ΔHR). For work conditions where RPE is greater than 5, the physiological demands of those tasks need to be reduced through appropriate interventions.
- Make a qualitative judgement on the biomechanical or physiological effect.
- Identify intervention strategies and rank order them based on (i) effect on body segments stressed, and (ii) main effects as observed from ANOVA analysis of 'muscle effort' & 'change in heart rate' (Rajesh 2015). The order of the intervention priority is to focus on the biomechanical effect and then the physiological effect.

3.3 Results and Discussion

3.3.1 *Biomechanical, Physiological and Perceptual Responses*

Varying degree of biomechanical and physiological stresses can be observed over the 22 worksystem settings. Table 3.1 shows the correlation between biomechanical, physiological and perceptual responses. The Spearman correlation coefficient between ME_{WB} and BPDFS is 0.74 and that between ME_{WB} and RPE is 0.72. Overall there are strong correlations between the biomechanical measure and the outcome measures (Table 3.1). The biomechanical measure ME_{WB} is moderately correlated ($\rho = 0.46$) with the physiological measure ΔHR. It is found that there is a moderate correlation ($\rho = 0.55$) for ME_{WB} and 'net energy expenditure'. Total energy expenditure (EEt) estimated has low correlation with RPE while energy expenditure rate (EEr), and net change in heart rate (ΔHR) has strong correlations with RPE. The unusually low correlation observed between EEt and RPE (0.23) could be due to the difference in biomechanical and physiological loads during the task. The key task variables that contribute to this effect are 'intensity' and 'duration'. In low 'intensity' task there are significant proportions of no load conditions, i.e. standing and walking, which does not affect exertion perceived by the worker. The rate function (EEr) appears to better represent exertion than total energy (EEt).

3.3.2 *Interventions Based on Muscle Effort*

A number of generic design solutions were examined and discussed with the Store head, Engineers and the workers involved. Each of the designs was modelled in CATIA software (Version 5, Dassault Systems and France). The choice of intervention was based on examining the biomechanical and physiological demands placed. Some of the intervention strategies that could be used to reduce MMH exposure are listed below. Few examples are shown in Figs. 3.1 and 3.2.

Table 3.1 Correlation between biomechanical, physiological and psychophysical responses for combined MMH task

	dHR	EEr	EEt	BPDFS	RPE
ME_{WB}	0.46	0.47	0.55	0.74	0.72
dHR		0.84	–	0.66	0.73
EEr			0.33	0.58	0.73
EEt				0.33	0.23
BPDFS					0.83

ME_{WB}—Muscle effort; dHR—Net change in heart rate; EEr—Energy expenditure rate; EEt—Total energy expenditure; BPDFS—Discomfort rating; RPE—Exertion level

(a) (b) (c)

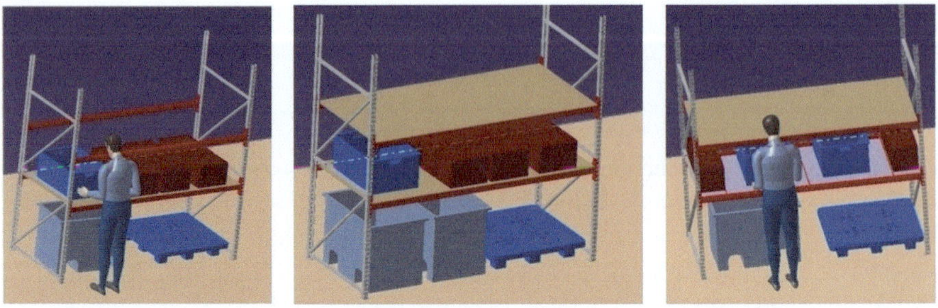

Fig. 3.1 **a** Single-shelf rack configuration. **b** Height-adjustable rack. **c** Rack shelf with rollers

(a) (b)

Fig. 3.2 **a** Redesign slopped rack with height-adjustable scissor table. **b** An example of slopped rack in the plant studied

- Clear surface.
- Height-adjustable pallet truck and single-shelf rack.
- Redesigned height-adjustable rack with height-adjustable pallet truck or scissor truck.
- Slopped rack with height-adjustable scissor table.
- Safe operating procedure for pushing and pulling loads.
- Safe operating procedure for MMH cycle.

A higher BPDFS indicates greater stress on the body segment and needs priority. Worksystem settings T3, T5, T8, T9, T10, T12, T13, T15, T17 and T21 have been identified as requiring intervention. From a whole-body exertion perspective, T9 and T10 appear to have low severity (BPDFS = 0.62 for T9 and BPDFS = 0.71 for T10). Examination of the trials identified as 'High' exertion task revealed that except for T10 all trials had 'Weight' settings of 20 kg (lifting/carrying) and 700 kg (pushing/pulling). T2 and T20 with same 'Weight' settings were not classified as 'High'. For T2, it was primarily because of two men undertaking the task, and for T20 it was due to multiple factors (i.e. low pace, short cycle, favourable height, no carrying, short distance and clear path).

3.3.3 *Interventions Based on Physiological Response*

The minimum physiological exertion was observed for T16 (mean net EEr = 3.0 kcal/min, SD = 0.03) and maximum for T21 (mean net EEr = 11.3 kcal/min, SD = 0.93). T3, T5, T6, T8, T10, T11, T13, T15, T17, T21 and T22 were considered 'heavy' or 'very heavy work' based on the classification of work using energy expenditure and heart rate guidelines (Kroemer et al. 2010). For these MMH tasks, the mean predicted energy expenditures are 5.5–11.3 kcal/min and mean heart rates are 120–151 bpm. Among the eleven MMH tasks identified only T5 was done at slow pace, and the mean EEr is 6.7 kcal/min and mean HR is 120 bpm. T5 consists of (i) loading component, i.e. cycle 1(1): loading 30 boxes of 20 kg located at waist–shoulder height on rack to pallet truck, by one man, at a rate of <5 lifts/min; (ii) moving component, i.e. cycle 1(2): pushing for a distance of 72 m at a speed of <0.3 m/s; and (iii) unloading component, i.e. cycle 1 (3): unloading 30 boxes of 20 kg located from the pallet truck back to rack at waist–shoulder, by one man, at a rate of <5 lifts/min; repeat the whole cycle again. The mean duration of the task is 53.3 min. The other exception was T1, in which the mean heart rate was observed as 119.8 bpm, close to the 120 bpm and be classified as 'heavy' (Kroemer et al. 2010). Hence, the work conditions simulated in this study at high frequency cause high stress on the physiological system. The MMH tasks done at a slower pace (<5 lift/min and <0.3 m/s) produced heart rates <120 bpm and energy expenditures <5 kcal/min. From a physiological perspective, the work conditions that are less stressful are T2, T4, T7, T9, T12, T14, T16, T18, T19 and T20. For the MMH tasks classified as 'heavy' or 'very heavy work' (based on EEr), the mean ratings of perceived exertion range from 4 to 7.5. This indicates a 'somewhat strong' to 'very strong' rating of perceived exertion level. The physiological demands of those tasks, therefore, need to be reduced through appropriate interventions. The simulated worksystem settings where the RPE is above 5 is considered for specific interventions. Those tasks are T3, T5, T13, T15, T17 and T21. Figure 3.3 shows the working heart rate for six 'heavy' MMH trials. It can be seen that all the six 'heavy' MMH tasks have the mean heart rate of above 120 bpm. The high exertion combined MMH tasks are compared with acceptable MMH tasks so that interventions can be suggested.

3.3.4 *Interventions Based on Biomechanical and Physiological Responses*

Table 3.2 summarizes the classification of trails based on biomechanical analysis and physiological analysis. From biomechanical analysis T3, T5, T8, T9, T10, T12, T13, T15, T17 and T21 were identified as a task requiring intervention (Sect. 3.2). A classification of task based on mean heart rate indicated that T1, T3, T5, T8, T10, T11, T13, T15, T17, T21 and T22 were 'heavy work'. Except for T5 all the tasks

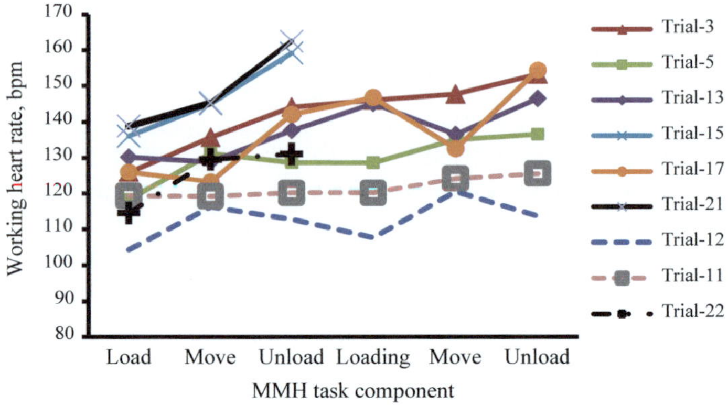

Fig. 3.3 Working heart rate for the high exertion tasks along each of the MMH task components

Table 3.2 Summary of classification of tasks using biomechanical and physiological responses

Trial	ME_{WB}	HR	ΔHR	RPE	BPDFS@	HR@	RPE@
1	239	123	32	3.0		H[1]	
2	806	109	23	3.3			
3	958	146	56	7.5	H	H	H
4	301	99	16	2.3			
5	1219	125	32	6.0	H	H	H
6	178	118	33	4.3			
7	263	107	24	2.3			
8	619	121	33	4.0	H	H[1]	
9	603	105	19	3.0	H[1]		
10	284	139	44	4.8	H[1]	H[1]	
11	599	125	35	4.1		H[1]	
12	1593	110	20	4.0	H		
13	1913	134	46	6.9	H	H	H
14	156	98	15	2.0			
15	691	142	58	6.3	H	H	H
16	381	94	13	3.1			
17	1560	136	48	6.9	H	H	H
18	496	110	20	4.0			
19	378	99	12	3.3			
20	481	118	26	4.3			
21	798	149	63	5.8	H	H	H
22	627	131	43	4.4		H[1]	

Note Trial—Worksystem settings; ME_{WB}—Muscle effort; BPDFS@—Classification based on BPDFS; ΔHR—Mean change in heart rate; HR—Mean heart rate; HR@—Classification based on HR; RPE—Exertion level; RPE@—Classification based on RPE; H[1] denotes classification difference arising from the physiological response (HR, RPE) and biomechanical response (BPDS, BPDFS) were high frequency (f >5/min) and high-paced tasks (speed >0.3 m/s). There is a classification difference arising from the physiological response (HR, RPE) and biomechanical response (BPDS, BPDFS) in the case of T1, T8, T9, T10, T11 and T22, and interventions needs to be undertaken only after careful examination. For these cases, HR is high but RPE <5. T1 is a low-weight setting task done by two men, and the task duration is short. For this condition, RPE classification is better, and hence intervention is not required. T8 is a high-weight setting task has high-weight setting condition done by two men with short duration. Even though the task is fast paced the physiological response is not very high. The primary reasons are (i) the task is done by two men, and (ii) it is a short cycle task. The strategy, in this case, is to reduce the weight setting rather than reducing the frequency. T9 is a high-weight setting task done by a single worker, and duration is short. The biomechanical demand for the task is greater than the physiological demand on the worker. An examination of the BPDFS (mean = 0.62, SD = 0.13) indicates that the discomfort severity is low and equivalent to low exertion tasks. Therefore, an intervention is not needed for T9. T10 is a low-weight setting task but is fast paced, and duration is long. From biomechanical perspective the BPDFS (mean = 0.77, SD = 0.2) indicates that the discomfort severity is low and equivalent to low exertion tasks, but BPDS for the lower back is 'moderate' in one case. This may due to long exposure or due to the strength profile of this worker. In this case, the HR (mean = 139, SD = 5.6) and RPE (mean = 4.8, SD = 0.5) indicate that the physiological response is high. Therefore, there is a justification to undertake intervention in T10. The strategies for this case are (i) reduce pace setting, (ii) make use two men or (iii) reduce task duration. The choice of the strategy would be decided by the main effects of the task variables. T22 is a low-weight setting task done by two men, fast paced for the long duration with long path and constrained path. In this case, workers do not face discomfort but physiological response appears to be high (HR = 131, RPE <5). It appears that the effect of fatigue has not set in, and hence the worker perceives lower exertion (i.e. RPE <5). Hence, it was decided not to undertake intervention.

3.3.5 Intervention for T5

The intervention procedure described in Sect. 2.2 is used to arrive at the feasible solutions. Details of the interventions for T3, T5, T10, T13, T15, T17 and T21 can be found in Rajesh (2015). Intervention for T5 is described here.

(a) **Task description**: The task consists of (i) loading component, i.e. cycle 1(1): loading 30 boxes of 20 kg each, located at waist–shoulder height on rack to pallet truck, by one man, at a rate of <5 lifts/min; (ii) moving component, i.e. cycle 1(2): pushing for a distance of 64 m at a speed of <0.3 m/s; and (iii) unloading component, i.e. cycle 1(3): unloading 30 boxes of 20 kg each, located from the pallet truck back to rack at waist–shoulder, by one man, at a rate of <5 lifts/min; repeat the whole cycle again. The mean duration of the task is 53.3 min.

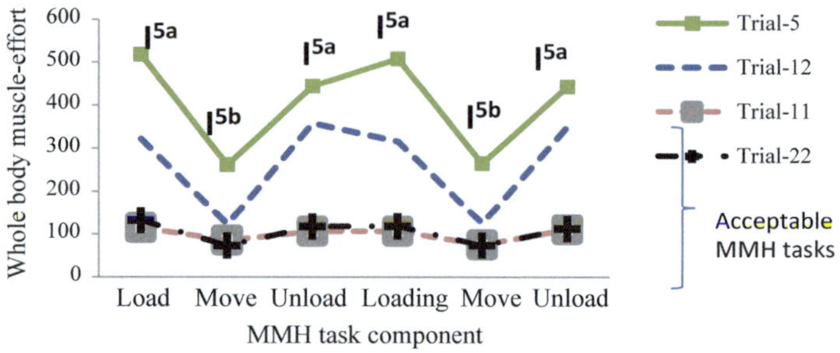

Fig. 3.4 Whole-body muscle effort for T5 and its intervention. *Note* I^{5a} denotes intervention (**a**) for task T5 and I^{5b} denotes intervention (**b**) for task T5

(b) **Issues identified**: High ME is observed during all MMH components over two cycles, i.e. loading, moving and unloading (Fig. 3.4). The BPDFS is high (mean = 1.3, SD = 0.1) and RPE is high (mean 6.0, SD = 0.0). Since both ME and RPE are high, both the biomechanical and physiological effects are significant, and call for examining the biomechanical and physiological responses of the material handler. As the pace of the task is <5/min, the high RPE response from the material handler for this combined MMH task can be attributed to the biomechanical stress rather than the physiological stress. An intervention from a biomechanical angle can be effective. The biomechanical response is focussed first and then the physiological responses towards deciding interventions.

(c) **Intervention**: Fig. 3.4 shows the variation of muscle effort over all the MMH components. On observing the BPDFS values it can be found that priority of intervention is to address the strain on the LE followed by that on LB or UE. The question now is on the solutions through which this can be achieved. Some of the solutions in Sect. 3.2 can be examined towards achieving a lower muscle effort demand.

Intervention I^{5a}—'*Making the task a team task*'. This would reduce the ME from 2438.8 to 1659.1. As the ME is still high, intervention I^{5b}—'*Clearing the aisle would reduce the ME during pushing-pulling task*' needs to be implemented. This will reduce the ME from 1659.1 to 1470.7 (overall 40% reduction). The predicted heart rate is now 109 (9% reduction) which is below the physiological limit of 120 bpm (see Rajesh 2015). The predicted biomechanical and physiological responses are in fact now lower than T12.

3.4 Conclusions

This paper highlights few interventions for an MMH work system in a bearing manufacturing plant in India through an experimental study. The experimental study reveals that the workers are faced with both biomechanical and physiological strains. The design solutions suggested above have the opportunity to address the whole-body exertions faced by the material handlers in an MMH work system. Some of the design recommendations include the following: reduce the weights of boxes; use height-adjustable scissor table or use height-adjustable pallet truck; use height-adjustable rack; avoid use of unfavourable storage heights; reduce push–pull distance; make clear aisle for pushing and pulling of material handling equipment; reduce the loading–unloading pace; provide additional worker for sharing the load; use better designed MHE; place MHE equipment appropriately to minimize postural manoeuvring and reduce the distance during carrying; and use rest breaks for high-paced repetitive tasks.

The practical implication of the study is that it highlights a few work system characteristics that produce less biomechanical and physiological stresses. The recommended interventions are generic and can be used in any MMH work system where human effort is dominant, and level of mechanization is low. Within Indian perspective, these solutions are relevant for a large number of small- and medium-scale industries.

The values obtained from this simulation are only the first approximation and need to be verified through actual implementation of the intervention. To ascertain whether the redesigned work system or task is safe (i.e. acceptable risk of injury) is farfetched as muscle effort design guidelines have not yet been established. There is a scope for developing an integrated measure that accounts for biomechanical and physiological stresses.

Acknowledgements The authors appreciate the efforts provided by Mr. Sukanta Chanda and Dr. O.B. Krishna pertaining to the experimental study. The authors thank Rakesh V.B. and Saran R. for their inputs on designs.

References

Ayoub MM (1998) A 2-D simulation model for lifting activities. Comput Ind Eng 35(3–4): 619–622

Borg G (2004) Scaling experiences during work: perceived exertion and difficulty. Handb Hum Factors Ergon Meth 11(1–11):7 (CRC Press)

Chaffin DB (2005) Improving digital human modeling for proactive ergonomics in design. Ergonomics 48(5):478–491

Ciriello VM (2005) The effects of box size, vertical distance, and height on lowering tasks for female industrial workers. Int J Ind Ergon 35(9):857–863

Ciriello VM (2007) The effects of container size, frequency and extended horizontal reach on maximum acceptable weights of lifting for female industrial workers. Appl Ergon 38(1):1–5

Garg A, Chaffin DB, Herrin GD (1978) Prediction of metabolic rates for manual materials handling jobs. Am Ind Hyg Assoc J 39(8):661–674

Hu B, Ma L, Zhang W, Salvendy G, Chablat D, Bennis F (2011) Predicting real-world ergonomic measurements by simulation in a virtual environment. Int J Ind Ergon 41(1):64–71

Kroemer KHE, Kroemer HJ, Kroemer-Elbert KE (2010) Engineering physiology—bases of human factors engineering/ergonomics, 4th edn. Springer, New York

Kuijer P, Frins-Dresen M, de Looze M, Visser B, vab der Beek A (2000) Work situation and physical workload of refuse collectors in three different time periods. Int J Ind Ergon 26: 509–519

Olendorf MR, Drury CG (2001) Postural discomfort and perceived exertion in standardized box-holding postures. Ergonomics 44(15):1341–1367

Perez J, Neumann WP (2015) Ergonomists' and engineers' views on the utility of virtual human factors tools. Hum Factors Ergon Manuf Serv Ind 25(3):279–293

Rajesh R (2015) Ergonomic modelling of combined manual material handling tasks. Ph.D. Thesis, Indian Institute of Technology, Kharagpur

Rajesh R, Maiti J (2011) Implication of container clearance on materials handling by pallet truck. International Ergonomics Conference HWWE 2011, IIT Madras

Rajesh R, Maiti J, Samanta B (2014) Application of the cube model for biomechanical exposure assessment of combined manual material handling tasks in a manufacturing plant in India. IIE Trans Occup Ergon Hum Factors 2:39–51

St-Vincent M, Denis D, Imbeau D, Laberge M (2005) Work factors affecting manual materials handling in a warehouse superstore. Int J Ind Ergon 35(1):33–46

Village J, Searcy C, Salustri F, Patrick Neumann W (2015) Design for human factors (DfHF): a grounded theory for integrating human factors into production design processes. Ergonomics 58(9):1529–1546

Wells R, Mathiassen SE, Medbo L, Winkel J (2007) Time: a key issue for musculoskeletal health and manufacturing. Appl Ergon 38(6):733–744

Young JG, Lin J, Chang C, McGorry RW (2013) The natural angle between the hand and handle and the effect of handle orientation on wrist radial/ulnar deviation during maximal push exertions. Ergonomics 56(4):682–691

Yun L, Xiaoqi Z, QianXiang Z (2013) Study on fatigue analysis and evaluation method of ergonomic virtual human. In World Congress on Medical Physics and Biomedical Engineering May 26–31, 2012, Beijing, China (pp 2011–2014), Springer, Berlin, Heidelberg

Chapter 4
Computer-Assisted Industrial Ergonomics: A Review

Harish Kumar Sharma, Piyush Singhal and Pankaj Sonia

Abstract Ergonomics involves a multidisciplinary approach that requires knowledge from several branches of engineering and sciences. The paper proposes a literature survey on computer-integrated techniques for the identification of industrial ergonomic factors. The motivation behind the paper is to review the applications of software packages in research papers of different authors to design the industrial workplace. The increasing complexity of industrial ergonomics is continuously evolving new challenges for the researchers. Ergonomics aims at safety, health, and efficiency of the worker. Therefore, the accessible evaluation instruments and strategies should risk chance variables that are in charge of musculoskeletal issue. It is important to check just the accessibility of techniques and apparatuses that consider the most up and coming. The article will subsequently break down and evaluate the powerful modeling and simulation software packages and programs. In the light of these observations, the authors find diverse applications of various programs. The computerized human model, likewise manikin, is a key some portion of human displaying programming. Digital human modeling (DHM) is an advanced representation of the human embedded into a virtual environment to encourage expectation of well-being as well as execution. The examination of the literature uncovers qualities of the methodologies, which have been connected by a few researchers to enhance the viability of people in their working environment. General, perception considers costs, capacity, and versatility for the best option to the needs for establishing intervention. This review summarizes the results of analyzing 83 papers, articles, and books. From this literature review, we ended up recognizing the role of software in the implementation of several software packages on workplace environment for the safety and efficacy of the worker.

Keywords Ergonomics · Industrial workplace · Modeling and simulation · Digital human modeling

---

H. Kumar Sharma (✉) · P. Singhal · P. Sonia
Department of Mechanical Engineering, GLA University, Mathura, Uttar Pradesh, India
e-mail: harish.shrma@gla.ac.in

P.K. Ray and J. Maiti (eds.), *Ergonomic Design of Products and Worksystems—21st Century Perspectives of Asia*, Managing the Asian Century,
https://doi.org/10.1007/978-981-10-5457-0_4

4.1 Introduction

A review of recent computer-aided ergonomics in last two decades reveals new concepts and approaches proposed. The purpose of the review is to enlist and compare available methods in occupational ergonomics. In the light of these observations, the authors find diverse applications of various programs. This review summarizes the results of analyzing 83 papers, articles, and books. Both nationally and internationally available scholarly journals, papers, articles, and publications identified with the point of this review have been considered by the authors. This contribution shows how computer-aided investigation can be utilized to increase particular signs on ergonomic considerations for the industrial workplace to be designed to meet the abilities. The first step for the ergonomic analysis is to capture body postures and movements and analyze using standard analysis methods with the help of software and verified by biomechanical approaches.

Human modeling tools are utilized to simulate the individual physically interfacing with the product and the environment. This permits a quantitative appraisal of openness and visibility of products, as asked for in this particular application, but, for instance, posture analysis and tasks evaluation can be done with a similar approach (Regazzoni et al. 2015).

Diverse digital models are today included into several commercial software toolkits, from SAFEWORK (by Dassault Systèmes) to Jack (by Siemens) accessible on Siemens/Tecnomatix products, until particular software for specific applications (SANTOS, 3DSSPP, or Anybody Modeling System) (Chaffin 2007). Human models can likewise be combined with multi-body and finite elements' analyses to get quantitative data on human–object interaction (Volpe et al. 2015).

Next sections describe different computer-aided tools and their modules of modeling, simulation, and analysis for industrial workplace design. In the last section, the conclusion reports contributions of the several authors.

4.2 Digital Human Modeling

Workstation assessments can be finished in two ways. The primary technique is the customary manual approach in which modern designers, qualified in ergonomics, have utilized physical models to perform general safety and ergonomic evaluations. The second approach, using virtual people, permits such assessment to be made utilizing 3D models of the workstations that have been created on CAD/CAM frameworks. This can be completed much sooner than physical models need be developed, packing the design life cycle.

Digital human modeling (DHM) is an advanced representation of the human embedded into virtual environment to encourage expectation of well-being as well as execution (Das and Sengupta 1996). DHM is proposed to bolster and encourage the new item and process plan, and empowers specialists of different controls to join ergonomics science and human components building standards in the working environment in which outlines prepare. Utilizing sensible advanced human as a part of PC recreation is a technique to take the early thought of ergonomics issues in the plan and to diminish the outline process duration and cost (Kushwaha and Prasad 2016). The computerized human model, likewise manikin, is a key some portion of human displaying programming. As indicated by Badler (Kushwaha and Prasad 2016), manikins can be portrayed for the most of the parts along five measurements: appearance (two-dimensional design, three-dimensional wireframe, and so on.), capacity (toon, skeleton with joints, active model, drive limits, and so on.), time reaction (disconnected reproduction, intuitive control, and so on.), self-rule (collaborating, deciding, and so on.), and distinction (sex and age, anthropometric information, and so forth.).

So as to utilize DHMs in a methodologically important manner, it is exceptionally useful to continue as indicated by a standardized approach.

These days, the utilization of the digital human modeling (DHM) software incorporated into the virtual manufacturing software (VMS) is the basic practice in numerous businesses, particularly those delivering complex items (e.g., vehicles and flying machines) or working with the complex creation and support processes (Chryssolouris et al. 2001).

A few commercial accessible DHM instruments have as of now been produced and incorporated into computer-aided design (CAD) tools, for example, Jack, 3DSSPP, RAMSIS, AnyBody, Santos, Human Builder, and so forth. DHM and virtual human simulations (e.g., 3DSSPP, EAI Jack, RAMSIS) have been made to encourage ergonomic assessments (Ma et al. 2009).

Raghunathan et al. (2013) manufactured a computerized human model in CATIA V5 for current worksystem. For the computerized human model anthropometric information of 75th percentile, Indian populace was utilized. At long last, the unsafe worksystem, area was overhauled ergonomically.

Marcos et al. (2006) go for diminishing the push of the therapeutic staff amid laparoscopic operations all the while expanding the security and productivity of a coordinated operation room. To this end, the authors develop a simulator by integrating the CAD software (CATIA) and the simulation software (RAMSIS).

Nishanth et al. (2015) considered an assembly workstation in a stator assembly area. Distinctive parts of pump are assembled by manual process. Assembly process comprises parts of various sizes and weights, inserted into the pump packaging. Anthropometry information are utilized for building manikin as a part of CATIA programming. The manikin is produced utilizing CATIA programming. Sánchez-Lite and García-García (2012) utilized CATIA V5 for the 3D CAD outline of the item and creation assets. The study is done for a little organization in the metal part with nine laborers on its staff.

Human builder is a part of the CAD application DELMIA by Dassault Systemes that gives an arrangement of instruments to look at items and procedures. It is additionally accessible in CATIA V5. Some DHM tools can figure the biomechanical traits (e.g., Anybody Modeling System, 3DSSPP) and anticipate physical weariness and potential issue chance. These logical apparatuses can be utilized to recognize and alleviate ergonomic issues of a composed item, workstation, or occupation to advance human considerations and ensure the clients, during an early phase of design (Mukhopadhyay et al. 2012).

4.3 Simulation Tools

The principle elements of the virtual human simulation tools are posture investigation and posture prediction. These devices are fit for deciding the workspace of virtual human, surveying the visibility, availability of an operation, and assessing posture. There are a few immersive virtual simulation and human modeling [IC. IDO by ESI Group (IC.IDO 2016)]. Such a setup permits fantastic stereoscopic and immersive virtual simulation to replicate virtual objects into a 1:1 scale, make exceptionally sensible simulations, and approve static and element practices during preparatory design. The substantial volume shown also supports collaborative design activities. To the extent advanced taunt ups creation, the 3D models of the mechanical plant and the workstations can be acknowledged by utilizing diverse 3D CAD programming (e.g., CATIA V5, JT, PLMXML, Pro/Engineer, SolidWorks, and so on.) and imported straightforwardly into IC.IDO programming. After that, the intelligent scene is made inside IC.IDO by characterizing the "touchable" items and the ecological imperatives (e.g., impacts, gravity, and so on.). Various scenes can be made with various properties. The client can associate straightforwardly with the virtual protests by Wiimote gadgets, and what is more he/she can be connected to a symbol that repeats his/her developments continuously. It can be acknowledged by interfacing a few focuses or gadgets, followed by the Vicon framework, with the symbol joints. For example, the Wiimotes can be connected to the symbol's hands or the took care of gadgets (i.e., screwier, gage, and so on.) while markers altered on the client's arms, legs, wrists, and feet can be connected to a similar symbol's body parts (Gill and Ruddle 1998).

Margherita Peruzzini et al. (2017) examinations grinding operation in the workstation. CATIA and DELMIA toolboxes were utilized to show and recreate the plant format and the workstation configuration highlights. Consequently, the virtual scenes were set up in IC.IDO considering the genuine grouping of activities and connection capacities with the workstation things (device, pipe, and so forth.). At long last, the client (simulating the real worker) was furnished with followed stereo glasses and devices (i.e., Wiimotes), which speak to take care of apparatuses, and connected with the virtual puppet in IC.IDO. Along these lines, the client can specifically communicate with the virtual scene, as indicated by the genuine creation requirements and occasions, while the virtual puppet moves as per his/her positions. In general, the human movement can be simulated and broke down in light of the workspace data, virtual human quality data, and different angles. In any case, there are still a few impediments in the current virtual human recreation instruments.

Rajan et al. (1999) built up a virtual reality-based environment JIGPRO for the examination of product assembly and jig design. 3D CAD models of assembly product, jig, and a virtual hand were foreign made into JIGPRO for get together process simulation and openness investigation. The fundamental reason for existing was to examine availability amid gathering and to lessen the danger of musculoskeletal injuries.

4.4 Analysis Tools

There are numerous software packages that have the usefulness to perform ergonomic appraisal including ERGO, SAMMIE, Jack, and so forth. Some DHM tools can ascertain the biomechanical properties (e.g., Anybody Modeling System, 3DSSPP) and anticipate physical fatigue and potential issue hazard. These investigative tools can be utilized to distinguish and moderate ergonomic issues of a designed product, workstation, or occupation keeping in mind the end goal to advance human considerations and protect the workers, amid an early phase of design (Mukhopadhyay et al. 2012).

DHM simulation frameworks, for example, Jack and RAMSIS, add to the productivity of item configuration prepare. These frameworks have been used as a viable outline device to envision the communication of a human and workstation framework. Some fatigue models have been consolidated into some virtual human devices to foresee the variable physical quality. For instance, Wexler's fatigue model (Ding et al. 2000) has been coordinated into Santos (Vignes 2004), and Giat's fatigue model (Giat et al. 1993) has been incorporated in light of Hill's muscle shown (Hill 1938) in the PC recreation by Komura et al. (2000).

Also, a few appraisals in those devices give lists produced by customary assessment strategies (e.g., Rapid upper appendage evaluation (RULA)). Due to the discontinuous recording methodology of the traditional stance investigation techniques, the assessment result cannot break down fatigue impact in detail (Komura et al. 2000).

4.4.1 3DSSPP (3D Static Strength Prediction Program)

The 3DSSPP was created at the College of Michigan, School of Building. This program can be utilized as a part of dissecting manual materials—taking care of undertakings. Ergonomists, architects, advisors, and scientists may utilize the product to assess and plan occupations. This program takes into account clients to include anthropometric information, and acquire the strengths and minutes registered by the program, instead of manual estimation. Moreover, the program likewise joins the National Institute of Occupational Safety (NIOSH) lifting information and other extra reports to recognize dangers connected with a specific undertaking. The 3DSSPP Static Quality Forecast Program Variant 6.0.3 Client Manual (2010) states that "Populace mean qualities… are registered from exact mean quality conditions".

4.4.2 Jack

Jack empowers you to size human models to match laborer populaces, and additionally test plans for numerous components, including damage chance, client comfort, approach, viewable pathway, vitality consumption, exhaustion limits, and other imperative human parameters. This distinction might be ascribed to Jack which gives more adaptability than 3DSSPP to control the body joints. It is less demanding to control and pivot the joints on Jack than 3DSSPP. The Jack programming was produced by the PC Design Research facility of the College of Pennsylvania. Quantitative assessment apparatuses are recorded beneath and in Jack they are assembled in a different extra module called Task Analysis Toolkit (TAT).

- Lower Back Analysis (LBA): it decides the strengths following up on the low back by assessing stances and load conditions.
- Static Strength Prediction (SSP): it assesses the rate of the working populace ready to perform (as far as muscular static quality) an assignment.
- NIOSH: it ascertains a suggested weight constrain and a lifting record for a given manual lifting/bringing down the activity or a general list on account of numerous assignments.
- Manual Handling Limits (MHL): it evaluates errands that require conveying, lifting, bringing down, pushing, or pulling.

Notwithstanding the intrinsic adaptability of Jack, the product can be further altered to the requirements of the client using C/C++ and Lisp programming languages. This empowers users to incorporate with the package many capacities including investigation procedures and human execution models.

Badler et al. (1993) talk about numerous cases that show the potential customization of Jack. One customization could give visual criticism of data on the ergonomics of postures. Charts can be created to investigate the properties (comfort, work done, and vitality qualities) of the postures utilized as a part of specific tasks. Another customization maps quality and body push data specifically onto the body of the virtual human because of the relative torque stacks on the body utilizing a color scale. Zero torque stack demonstrates the applicable joint in white and most extreme torque stack hues the joint blue. On the off chance that the most extreme torque surpasses any joint quality then the joint is shaded red. This strategy is anything but difficult to decipher and it gives prompt criticism on the ergonomic condition of every stance amid element movement. This last customization would presumably be past the programming capacities of most users even after the Jack instructional classes. Programming providers could perform such customizations for their users.

In any case, even with this change the finishes of such an evaluation could in any case not be depended on if the client had little ergonomics preparing and encounter. There would be serious consequence results if mistakes had been made in the understanding of the package (Gill and Ruddle 1998).

Cocca et al. (2008) considered a movement comprisingan operation of visual and material assessment of hot-shaped glass items. Jack programming was chosen as it is supplied with criteria of examination.

Jayaram et al. (2006) exhibited a joining of virtual situations and quantitative ergonomic examination tool (Jack) into the real-time occupational ergonomic research. This examination permitted distinctive posture and assembly procedures to be analyzed in a more thorough way keeping in mind the end goal to distinguish issues in the assembly environment.

Kumar and Das (2012) performed act investigation in cast house work framework. The simulation of the current framework was performed in Jack 5.1 stage which empowers clients to position biomechanical exact advanced people of different sizes in virtual situations to allocate their undertakings and break down their exhibitions. From this study, 13 postures were distinguished which were embraced by throwing machine administrators to perform operations. The watched working postures were created in the computer simulation environment.

Bertoloni et al. (2012) considered software Jack (version 7.1) for Sonographers' Workplace Improvement.

Erika et al. (2013) in their article described the use of the Jack human model for the examination of vehicle departure by military staff taking after a mischance. In post-mishap circumstances, the vehicle state of mind may anticipate departure by the typical strategies. This paper presents a case study of the application of the Jack.

Sundin et al. (2004) depicted a contextual investigation of bus chassis assembly, which expected to enhance the effectiveness and ergonomics in the early design phase of products. 'Jack' was utilized for making the computer manikin and leading ergonomic examination as far as various working sequences and postures. The experience acquired for this situation concentrates on demonstrated that an ergonomics approach enhances the design of product and production, prompting to a superior final product and better assembly operations in the production framework.

Jinyan (Christy) Du et al. considered an ergonomic assessment which gathered the real task practitioner's motion at an automobile assembly line and utilized the movement information to drive the posture of the manikin. A few conventional ergonomic examination techniques inserted in Jack are utilized to assess the potential damage of the undertakings. The technique joins the benefits of both real task movement and computerized human simulation in a field study with experienced subjects. Every single collected data were dissected after the field tests.

4.4.3 Ergowatch

Ergowatch is a mechanized ergonomics package framework. The Ergowatch bundle incorporates NIOSH device which gives stack cutoff points to lifting and bringing down exercises.

The Anyone Demonstrating Framework is a product framework for reproducing the mechanics of the live human body working in a specific domain. Anyone has applications in the automotive industry, pharmaceutical, and sports examination. The product runs a recreation and figures the related mechanical properties including singular muscle powers, joint powers and minutes, digestion system, flexible vitality in ligaments, opposing muscle activities, and a great deal more. Anyone can likewise import information from Movement Catch frameworks.

Raghunathan et al. (2013) broke down an instance of an assembling plant toward lessening the physical strain for laborers undertaking material taking care of in a worksystem.

4.4.4 DELMIA

DELMIA is a package which offers a prepared, to utilize, par-based ergonomics arrangement which is utilized to assess and investigate human action in the working environment. Its elements incorporate a different vision window that shows the vision field from the puppet's point of view to survey visual impedance. Other key components incorporate the following:

- Lifting, bringing down, and conveying assignments utilizing the NIOSH 81 and 91 conditions;
- Push and draw undertakings utilizing the Ciriello and Snook (1999) conditions; and
- RULA for arm position appraisal, with the capacity to redo RULA determinations for biomechanical evaluation of the undertakings.

Sánchez-Lite and García-García (2012) utilized CATIA V5 for the 3D CAD design of the product and production resources and DELMIA HUMAN V5 for incorporation of the human element. The study is completed for a small organization in the metal division with nine laborers on its staff. An ordinary working day comprises eight hours. Inside its action, the organization incorporates different assembling procedures, for example, assembly, abrasive waterjet cutting, high definition plasma cutting, and profile rolling.

The utilization of DELMIA for the manual assembly process simulation has ended up being gainful to the item and work environment design. Be that as it may, for more complex posture simulations, the use of DELMIA requires more care with a specific end goal to prevent improbable results. This depends on encounters in programming control and the acknowledgment of various assembly tasks attempted.

4.4.5 AnyBody

AnyBody Displaying Framework is a product package for reenacting the mechanics of the live human body working in a work environment. AnyBody has applications in the automobile business, drug, the airplane business, sports investigation, explore, and even safeguard. The product runs a reproduction and computes the related mechanical properties including singular muscle strengths, joint powers and minutes, digestion system, versatile vitality in ligaments, opposing muscle activities, and significantly more (Singh 2010).

John Rasmussen et al. utilized the AnyBody Demonstrating Framework for computing the relating strong action for various extents for the improvement of a hand saw as for digestion system.

A model of the abdominal area working the saw is produced. The model involves 52 muscles, 5 fragments, and 6 joints.

4.4.6 Realistic Anthropological Mathematical System for Interior Comfort Simulation (RAMSIS)

The human model known as the acronym RAMSIS was initially created at the Institute of Ergonomics (TU München), Katholische Universität Eichstätt, and an examination relationship of automotive manufacturers. RAMSIS is for the most part utilized for product design. Particularly adjusted to the requirements of the automotive industry, it can be utilized as a solitary application or as a coordinated application in the CAD framework CATIA. It is additionally accessible as a generation situated usage in the PLM software Tecnomatix. The fundamental uses of RAMSIS are ergonomic studies in cockpits of vehicles, airships, and development apparatuses. The framework incorporates many modern databases with a few populaces of various ages, separation as indicated by percentiles, and incorporates the development of increasing speed. Notwithstanding the human models, which have the biggest appropriation, there are others that are additionally named ergonomic DHMs. Their dispersion among professionals is less critical or their scope of capacities is restricted.

"Anyman" is like Jack as the previous part of the Tecnomatix System. Indeed, even in the CAD framework CREO, once in the past known as PRO/Engineer Wildfire the human model called manikin is accessible. The human model actualized in DELMIA Process Engineer (DPE) is utilized for work put examinations with specific concentrate on financial time administration issues. A genuinely new, however, progressively critical DHM is the EMA by IMK car which incorporates generation ergonomics and process arranging in the view of MTM standard times (Schaub et al. 2012).

The survey of existing computerized human modeling tools has uncovered one constraint, i.e., the trouble to anticipate complex human postures and developments in an auspicious and practical way. A general pragmatic issue confronted is that users were unequipped for determining how a manikin of certain demographic and anthropometric attributes ought to be situated in the virtual environment, particularly when dynamic exercises or movements are included.

Some product offers the likelihood to build software applications inside the particular software by means of an Application Programming Interface (API). The API fills in as an interface between the product engineer and the product. The principle design is to get the capacities inside the product. This can be by means of different projects, for instance, between a CAD program and a simulation program; however, it can likewise be utilized to make macros inside a particular software (Hogberg et al. 2008).

4.5 Conclusion

The decision between the packages of ergonomic assessment will rely on the application concerned and the goals of the study. Generally, perception considers costs, capacity, and versatility for the best option to the needs for establishing intervention. This study comprised a comprehensive literature review of the last two decades regarding the application of several computer-aided methods and their possibilities in the industrial ergonomics. On the off chance that such programs are presented for full proactive assessment, the huge cost of amending design blunders after generation can be significantly decreased, notwithstanding the investment funds increased through customary ergonomic appraisal. Facilitate improvements in virtual human software and strategies for connection with virtual people will, in time, give a valuable instrument that permits all plan architects to make educated judgments about ergonomics, as a feature of the concurrent design process.

The utilization of these tools in the ergonomics simulation and analysis is costly since it has requesting prerequisites on computer software tools and hardware for the creation and representation. Also, so as to pick up the mix of CAD frameworks, a tedious information trade method is necessary. Hence, its application in ergonomics research is still constrained, particularly for those little and medium ventures despite the fact that they have a similar danger of ergonomic issues.

From this literature review, we ended up recognizing the role of software in the implementation of several methods on workplace environment for the safety and efficacy of the worker.

Acknowledgements The authors appreciatively recognize the support and collaboration given by the administration of GLA University and Head of Mechanical Engineering Department.

References

Badler NI, Phillips CB, Webber BL (1993) Simulating humans. Oxford University Press, Oxford vol 16, issue 1-1, p 79

Bertoloni E, Fenaroli M, Marciano F, Rossi D (2012) Sonographers, workplace improvement: ergonomics evaluation using modelling and simulation software. In: Proceedings of the 2012 international conference on industrial engineering and operations management Istanbul, Turkey

Chaffin DB (2007) Human motion simulation for vehicle and workplace design. Hum Factors Ergon Manuf 17(5):475–484

Chryssolouris G, Karabatsou V, Kapetanaki G (2001) Virtual reality and human simulation for manufacturing. In: Proceedings of the 34th international CIRP seminar on manufacturing systems, Athens, Greece, pp 393–398

Ciriello VM, Snook SH (1999) Survey of manual handling tasks. Int J Ind Ergon 23(3):149–156

Cocca P, Filippo M, Rossi D (2008) Assessment of biomechanical risk at work: practical approaches and tools. Acta of Bioengineering and Biomechanics 10(3):21–28

Das, B, Sengupta, A.K (1996) Industrial workstation design: A systematic ergonomics approach, Applied Ergonomics, 27(3):157–63

David GC (2005) Ergonomic methods for assessing exposure to risk factors for work-related musculoskeletal disorders. Occup Med 55:190–199

Ding J, Wexler AS, Binder-Macleod SA (2000) A predictive model of fatigue in human skeletal muscles. J Appl Physiol 89:1322–1332

Erika B, Budzik S, Reed M (2013) Modeling post-accident vehicle egress, 2nd International Digital human modeling symposium, Ann Arbor, Michigan University

Giat Y, Mizrahi J, Levy M (1993) A musculoskeletal model of the fatigue profiles of paralyzed quadriceps muscle under FES. IEEE Trans Biomech Eng 40(7):664–674

Gill S, Ruddle R (1998) Using virtual humans to solve real ergonomic design problems. In: Proceedings of the 1998 international conference on simulation, pp 223–229

Hill A (1938) The heat of shortening and the dynamic constants of muscle. Proceed Roy Soc London Ser B Biol Sci 126(843):136–195

Hogberg D, Backstrand G, Lamkull D, Hanson L, Ortengren r (2008) Industrial customisation of digital human modelling tools. Int J Serv Oper Inf 3(1):53–70

IC.IDO (2016) Accessed in April 2016, available at: https://www.esi-group.com/softwaresolutions/virtual-reality/icido

Jayaram U, Jayaram S, Shaikh I, Kim Y, Palmer C (2006) Introducing quantitative analysis methods into virtual environments for real-time and continuous ergonomic evaluations. Comput Ind 57(3):283–296

Komura T, Shinagawa Y, Kunii TL (2000) Creating and retargeting motion by the musculoskeletal human. Vis Comput 16(5):254–270

Kumar GS, Das A (2012) Analysis and ergonomic improvement of working postures in cast house work station using JACK modelling. Int J Hum Factors Model Simul 3(1):16–31

Kushwaha DK, Prasad VK (2016) Ergonomic assessment and workstation design of shipping crane cabin in steel industry. Int J Ind Ergon 52:29–39

Lin RT, Chan CC (2007) Effectiveness of workstation design on reducing musculoskeletal risk factors and symptoms among semiconductor fabrication room workers. Int J Ind Ergon 37:35–42

Ma L, Chablat D, Bennis F, Zhang W (2009) Anew simple dynamic fatigue muscle model and its validation. Int J Ind Ergon 39(1):211–220

Marcos P, Seitz T, Bubb H, Wichert A, Feussner H (2006) Computer simulation for ergonomic improvements in laparoscopic surgery. Appl Ergon 37:251–258

Mukhopadhyay S, Das SK, Chakraborty T (2012) Computer aided design in digital human modeling for human computer interaction in ergonomic assessment: a review. Int J Adv Comput Res 2(4). ISSN (print): 2249–7277; ISSN (online): 2277-7970)

Nishanth R, Muthukumar MV, Arivanantham A (2015) Ergonomic workplace evaluation for assessing occupational risks in multistage pump assembly. Int J Comput Appl 113(9) (0975 – 8887)

Peruzzini M, Carassai S, Pellicciari M, Andrisano AO (2017) Human-centred design of ergonomic workstations on interactive digital mock-ups. In book: Advances on mechanics, design engineering and manufacturing, pp 1187–1195

Rajan VN, Sivasubramanian K, Fernandez JE (1999) Accessibility and ergonomic analysis of assembly product and jig designs. Int J Ind Ergon 23(5–6):473–487

Rajesh R, Rakesh VB, Ramachandran S (2013) Ergonomic redesign of a material handling worksystem in a manufacturing plant. Int J Innovative Res Sci Eng Technol 2(1):582–588

Regazzoni D, Rizzi C, Colombo G (2015) The role of virtual ergonomic simulation to develop innovative human centered products. In: Digital human modeling. Applications in health, safety, ergonomics and risk management: ergonomics and health. Volume 9185 of the series Lecture Notes in Computer Science, pp 74–83

Sánchez-Lite A, García-García M (2012) Simulation and ergonomics approach for service and manufacturing process improvement. Key Eng Mater 502:121–125

Schaub KG, Mühlstedt J, Illmann B, Bauer S, Fritzsche L et al (2012) Ergonomic assessment of automotive assembly tasks with digital human modelling and the 'ergonomics assessment worksheet'(EAWS). Int J Hum Factors Model Simul 3:398–426

Singh LP (2010) Work Posture Assessment in Forging Industry: An Exploratory Study in INDIA. Int J of Adv Engg Tech, 1(3):358–66

Sundin A, Christmansson M, Larsson M (2004) A different perspective in participatory ergonomics in product development improves assembly work in the automotive industry. Int J Ind Ergon 33(1):1–14

Vignes RM (2004) Modeling muscle fatigue in digital humans. Master's thesis, Graduate College of The University of Iowa

Volpe Y, Governi L, Furferi R (2015) A computational model for early assessment of padded furniture comfort performance. Hum Factors Ergon Manuf 25(1):90–105

Chapter 5
Ergonomic Analysis of Work-Related Musculoskeletal Disorders for Radial Drilling Machine Operators

P. Ashok, G. Madhan Mohan and B. Siva Sankaran

Abstract Work-related musculoskeletal disorder (WMSD) is a common problem among workers in a radial drilling machine. The purpose of this study is to analyze the various awkward postures of workers working in a radial drilling machine and to identify WMSD. A questionnaire survey has been conducted in the industry among 36 male workers aged between 25 and 35 years. It contains information about their health condition, work methods, and body part discomfort. The average amount of working time per day is 9 h and work experience of 4–5 years. The muscles identified for investigation are sternocleidomastoid muscle (Neck), deltoid muscle (Shoulder), thoracolumbar muscle (Low back), brachioradialis, and flexor carpi-radialis (Forearm) muscles. Maximal voluntary contraction (MVC) of each muscle is identified and median frequency of the muscle is computed. The grip force dynamometer is used to measure the isometric muscular strength of the predominant arm. Maximum grip strength is observed to be at flexor 90°, elbow flexor 120°, and wrist 0°. The heart rate monitor was used to evaluate the physiological stress of different work postures of the workers in a radial drilling machine. Based on this study, it is concluded that risk of the worker is evaluated and compared with the modified posture by reducing the WMSD to improve the performance of workers and increase productivity.

Keywords Musculoskeletal disorder · Surface electromyography · Heart rate monitor · Radial drilling machine

5.1 Introduction

Work-related musculoskeletal disorder is one of the major problems for occupational workers (Wells et al. 1994). WMSDs can affect the quality of life, resulting in increased sick leaves or early retirement and can impose a major economic burden

P. Ashok (✉) · G. Madhan Mohan · B. Siva Sankaran
Department of Production Engineering, PSG College of Technology, Coimbatore, Tamil Nadu, India
e-mail: ashokdino@gmail.com; pak@pro.psgtech.ac.in

P.K. Ray and J. Maiti (eds.), *Ergonomic Design of Products and Worksystems—21st Century Perspectives of Asia*, Managing the Asian Century,
https://doi.org/10.1007/978-981-10-5457-0_5

(Cheng et al. 2013). The neck and shoulder region is the main discomfort muscle for the operators in radial drilling machine (Nordander et al. 2016). The operators of the machine occur the pain on brachii muscle activities (Forearm) (Ali et al. 2013). The operators also have awkward position while working and manual material handling (Ashish and Nimbarte 2014). The handheld dynamometer is to determine the worker's grip strength and the task performed in the machine. For modified workplace, the grip strength is to be evaluated (Andersen et al. 2014). The discomfort muscle parts are measured by surface electromyography by muscle frequency and contraction (Cashaback et al. 2013). The upper extremity exertions are mainly caused by MSD to modify the workplace and to improve the performance (Chowdhury et al. 2013). The individual work performance seems to be negative which will be the effects of obesity during manual material handling and musculoskeletal disorder (Colim et al. 2016). Low back pain is the major problem for the workers in the industry and fatigue assessment causes the pain (Curran et al. 2014). The muscle activation for the rest and working condition is to be compared with modified workplace to reduce the musculoskeletal disorder (Kim et al. 2014).

5.2 Methodology

5.2.1 *Questionnaire*

A questionnaire is a collection of worker information to identify the musculoskeletal disorder muscle (Heiden et al. 2013). It contains personal information, health condition, work method, body part discomfort (frequency and severity scale), and BORG RPE scale (overall exertion).

5.2.2 *Subjects*

Thirty-six male workers in industry participated in this study between the ages of 25 and 45 years. At the beginning of the experiment, anthropometric measurements of human height (cm) and shoulder height (cm) were made. Based on the height, the following three groups are for height (Table 5.1).

All subjects were initially fresher to the radial drilling operation. Subjects received training in handling and techniques from trainers. Each subject attends the demo class of the machine prior to testing, depending on the ability to perform the task effectively.

Table 5.1 Group of heights for gender

Height	Percentile	Male (cm)
Small	5–20	145–160
Medium	20–70	160–182
Large	70–95	182–195

5.2.3 *Work Method*

This study tested the postures, especially in radial drilling machine operators. The current study aimed to reduce the musculoskeletal disorder (MSD) that can be associated with these tasks. The tasks are normal posture, accuracy, push, and pull the machine. The tasks were selected because workers have extreme working postures, manual material handling, and impact movements (Tompa et al. 2010). The radial drilling machine is used for medium to large and heavy workpieces. It is used in manufacturing industries. It depends on manual activities such as manual material handling and work postures. The average amount of working time per day is 9 h and work experience of 4-5 years. The workers are given 2 break periods and one hour for lunch. The machine is to drill the number of components in a day.

Therefore, the operators have a cyclic work for the components. Due to poor working posture, pain occurred in the arm, neck, and back of the worker (Santos et al. 2016). Therefore, the experiment investigated the biomechanical response of different height workers.

5.3 Experimental Setup

Subjects are prepared for EMG recording. The detailed muscle activity is measured by Surface electromyography (sEMG) technique. From the data collection and survey results, the four painful muscles (Table 5.2) among workers at the time of working are sternocleidomastoid muscle (Neck), deltoid muscle (Shoulder), thoracolumbar muscle (Low Back), and brachioradialis muscle (Forearm). The subjects having excess hair shaved and the area have been rubbed with alcohol assure proper electrode–skin contact.

The signal quality was inspected visually before the measurement. When excessive noise was present, a small amount of gel was added to the electrodes or they were repositioned until signals with good quality were obtained. Two bipolar surface electrodes were placed on each muscle along its line of action. The integrated signal is fed into an analog to digital by delsys wireless system with respect to the workers muscle movement. The experiments have been carried out continuously in Existing workplace (Fig. 5.1) having poor working postures and modified workplace (Fig. 5.2). The results from the experiments have been compared and the modified workplace can be validated. The maximum hand grip strengths were performed with the participants in working position with a 120° elbow flexed, wrist 0°, and flexor 90° exerting a power grip in a hand dynamometer (Jamar plus + hand dynamometer). Each subject is to exert a maximum grip up to 10 s in both hands. Three trials are taken for each working position.

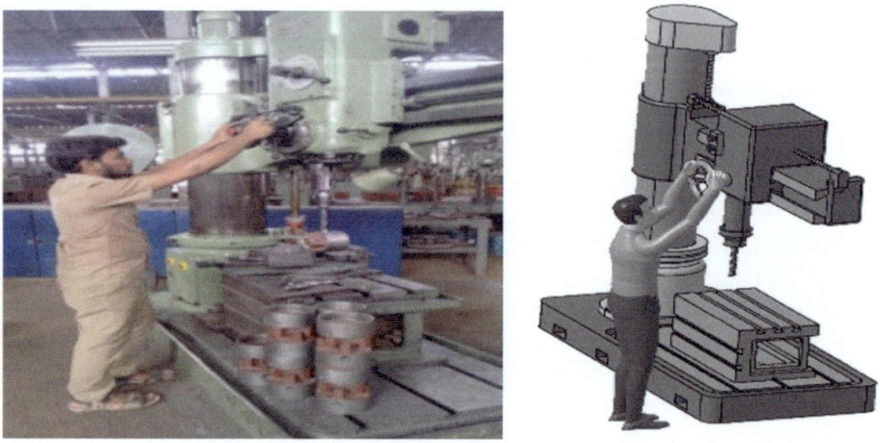

Fig. 5.1 Normal working posture of operator

Fig. 5.2 Modified work posture

5.4 Result

5.4.1 RULA Method

RULA is a rapid upper limb assessment tool to identify the score obtained by the posture. The following table shows that the workplace is awkward as compared to the modified posture (Table 5.3).

Table 5.2 Muscle location and names

S.no	Location	Muscle name
1	Neck	Sternocleidomastoid
2	Shoulder	Deltoid
3	Lowback	Thoracolumbar
4	Forearm	Brachioradialis

Table 5.3 RULA score

Workplace	Neck	Shoulder	Low back	Forearm	Final
Existing	4	4	3	2	7
New	1	2	2	1	3

Table 5.4 Grip strength data

	Armflexion 120°		Wrist 0°		Flexion 90°	
	Right	Left	Right	Left	Right	Left
Old	25	21.2	36.6	32	31.1	26
New	28.6	22	38	32.5	34.7	28.4

5.4.2 Grip Strength Dynamometer

The grip strength is the maximum exertion by the operator to control and operate the machine by hand feed wheel (Table 5.4).

5.4.3 Electromyography Analysis

The evaluation of the tasks such as normal posture, push, and pull the machine are measured by surface electromyography (sEMG). The muscles, deltoid muscle, sternocleidomastoid muscle, thoracolumbar muscle, and brachioradialis muscle, are evaluated by this technique. The data has been recorded from the different workplaces as per the work method. By comparing the measurements obtained from the different operators, the work-related musculoskeletal disorder can be reduced (Fig. 5.3 and Tables 5.5 and 5.6).

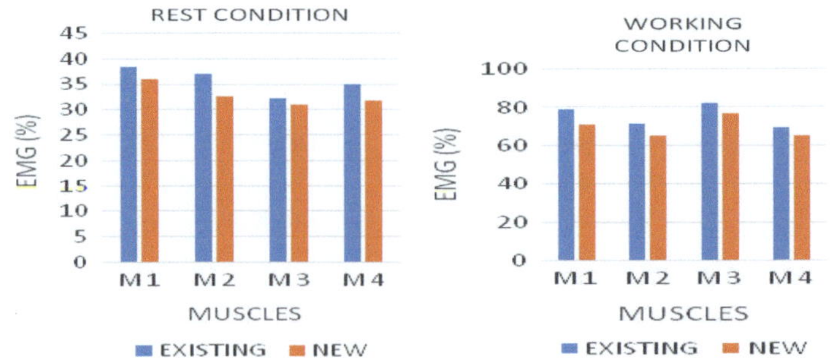

Fig. 5.3 EMG data when a different muscle is activated in rest and working condition (*M1*—Sternocleidomastoid muscle, *M2*—Deltoid muscle, *M3*—Thoracolumbar muscle, *M4*—Brachioradialis muscle)

Table 5.5 Muscle activation in rest condition

Rest condition				
	M1	M2	M3	M4
Existing	38.4	37	32.1	35
New	36	32.5	30.8	31.7

Table 5.6 Muscle activation in working condition

Working condition				
	M1	M2	M3	M4
Existing	79.2	71.6	82.1	69.2
New	71	65	77	65

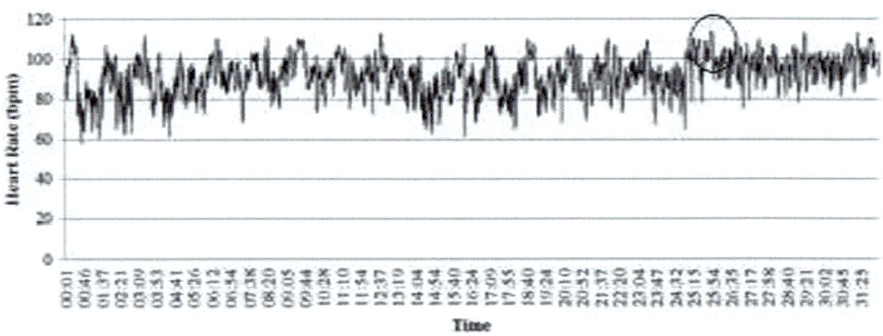

Fig. 5.4 Heart rate graph of worker

5.4.4 Heart Rate Monitor

Before operating the radial drilling machine, the operator's heart beat rate was in the range of 65-73 bpm (Fig. 5.4).

The maximum heart beat rate region is indicated in circle. Heart rate has been recorded for the worker who operates a radial drilling machine for a time period of 32 min. The age of the person whose heart rate has been monitored is 23 years. The average heart rate of the person monitored during the process is 91 bpm.

5.5 Conclusion

The reduction of musculoskeletal disorder is the important factor to increase the performance of the operator (Tullar et al. 2010). By suggesting the alternate posture to reduce the risk, the work method is simple as compared to the previous posture. The productivity and efficiency are also increased in the manufacturing sector (King et al. 2009).

References

Ali MA, Sundaraj K, Ahmad RB, Ahamed NU, Islam A (2013) Surface electromyography for assessing triceps brachii muscle activities: a literature review. Biocybern Biomed Eng 33:187–195

Andersen KS, Christensen BH, Samani A, Madeleine P (2014) Between-day reliability of a hand-held dynamometer and surface electromyography recordings during isometric submaximal contractions in different shoulder positions. J Electromyogr Kinesiol

Ashish D, Nimbarte AD (2014) Risk of neck musculoskeletal disorders among males and females in lifting exertions. Int J Ind Ergon 44:253–259

Cashaback JGA, Cluff T, Potvin JR (2013) Muscle fatigue and contraction intensity modulates the complexity of surface electromyography. J Electromyogr kinesiol 23:78–83

Cheng HYK et al (2013) Work-related musculoskeletal disorders and ergonomic risk factors in early intervention educators. Appl Ergon 44:134–141

Chowdhury SK, Nimbarte AD, Jaridi M, Creese RC (2013) Discrete wavelet transform analysis of surface electromyography for the fatigue assessment of neck and shoulder muscles. J Electromyogr Kinesiol 23:995–1003

Colim AS, Arezes PM, Flores P, Braga AC (2016) The effects of obesity on muscle activity during lifting tasks. Occup Saf Hyg

Curran M, Dankaerts W, Sullivan PO, Sullivan LO, Sullivan KO' (2014) The effect of a backrest and seatpan inclination on sitting discomfort and trunk muscle activation in subjects with extension-related low back pain. Ergon 57(5):733–743

Heiden B et al (2013) Association of age and physical job demands with musculoskeletal disorders in nurses. Appl Ergon 44:652–658

Kim CY, Choi JD, Kim SY, Ohc DW, Kim JK, Park JW (2014) Comparison between muscle activation measured by electromyography and muscle thickness measured using ultrasonography for effective muscle assessment. J Electromyogr kinesiol 24:614–620

King P et al (2009) Work-related musculoskeletal disorders and injuries: differences among older and younger occupational and physical therapists. J Occup Rehabil 19:274–283

Nordander C, Hansson GA, Ohlsson K, Arvidsson I, Balogh I, Str€omberg U, Rittner R, StaffanSkerfving (2016) Exposure response relationships for work-related neck and shoulder musculoskeletal disorders—Analyses of pooled uniform data sets. Appl Ergon 55:70–84

Santos J et al (2016) The influence of task design on upper limb muscles fatigue during low-load repetitive work: a systematic review. Int J Ind Ergon 52:78–91

Tompa E et al (2010) A systematic review of workplace ergonomic interventions with economic analyses. J Occup Rehabil 20:220–234

Tullar JM et al (2010) Occupational safety and health interventions to reduce musculoskeletal symptoms in the health care sector. J occup reliab 20:199–219

Wells R, Moore A, Potvin J, Norman R (1994) Assessment of risk factors of work related musculoskeletal disorders. Appl Ergon 25(3):157–164

Chapter 6
Ergonomic Evaluation of Bar Benders in Construction Industry

S. Drisya, B. B. Das and Ratri Parida

Abstract Construction industry is one of the risky industries in which the chances for developing musculoskeletal disorders or discomforts (MSDs) are very high among the workers. Based on the literature review, one of the most significant risk factors is awkward posture during the work tasks. In addition, not much reported research in detail is available for the causes of their musculoskeletal disorders or discomforts. Among the various categories of workers in the construction industry, a special and skilled workforce which is normally neglected is the "Bar Benders". This paper investigates the prevalence of symptoms of MSD through a questionnaire survey and also attempts to build a relation of working postures and MSD through postural assessment tools, among the bar benders. A questionnaire survey was carried out on 25 bar benders to assess the MSD and from the compilation of results, the shoulder, knee, and upper back were the body parts most commonly reported for work-related MSD. Further, with the help of postural assessment tools, i.e., REBA (Rapid Entire Body Assessment) and RULA (Rapid Upper Limb Assessment), the working postures of the selected bar benders were analyzed and the risk level of various postures (in total 25) were determined. From the results of the postural assessment, most of the working postures analyzed were at high or medium risk and recommended for changing the posture. Furthermore, remedial and corrective measures were suggested so as to improve the health and safety of the workers.

Keywords MSDs · Postural assessment · REBA · RULA

S. Drisya · B.B. Das
Department of Civil Engineering, National Institute of Technology Karnataka, Surathkal, Karnataka, India

R. Parida (✉)
School of General Management, National Institute of Construction Management and Research, Pune, Maharashtra, India
e-mail: ratriparida@gmail.com

R. Parida
Indian Institute of Management (IIM), Sambalpur, India

P.K. Ray and J. Maiti (eds.), *Ergonomic Design of Products and Worksystems—21st Century Perspectives of Asia*, Managing the Asian Century,
https://doi.org/10.1007/978-981-10-5457-0_6

6.1 Introduction

Construction industry faces many challenges and difficulties nowadays. It is one of the risky industries with a high level of injuries. Many construction work tasks are physically strenuous and which makes the health issues among construction workers is higher than other occupations (Rwamamara et al. 2010). The increasing number of work-related MSDs caused by awkward postures, excessive force, and repetitive motion; ergonomics has become a critical factor in workplace safety. The construction workers belong to unorganized work forces. Due to this, they are exploited since ages by not being provided proper shelter, pay, proper leaves and medical attention. The work task demands the workers to do repetitious movements, awkward postures and high force levels which lead to MSDs develop over a period of time and are not curable (Jaiswal and Veerkumar 2016). Among various types of work systems, construction is known for presenting a very high risk of MSDs to the workers as the workers are primarily engaged in a variety of manual material handling activities (Parida and Ray 2012). Musculoskeletal disorders (MSDs) are surrounded by the most common work-related problem in India. In an Indian construction industry, most of the work is still carried out manually hence issues of work-related musculoskeletal disorders in various body parts are prominent (Sain and Meena 2016; Parida and Ray 2011, 2015a, 2015b; Ray et al. 2015). Work-related musculoskeletal disorders (WMSDs) are described as a whole range of disorders which are not typically the result of acute or instantaneous event, but which is the result of a chronic development. Various ergonomic risk factors, including personal characteristics (for example physical limitations) as well as societal factors may contribute to the development of these disorders (Armstrong et al. 1993). Keeping in view the necessity of problems as stated, a questionnaire survey was conducted to study the socio-demographic profile and job characteristics, discomforts of musculoskeletal systems of the construction workers mainly focusing on bar bending workers. The qualitative study consisted of the postural assessment methods rapid upper limb assessment (RULA) and the rapid entire body assessment (REBA) (McAtamney and Corlett 1993; Hignett and McAtamney 2000). The RULA is a quick and systematic assessment that assesses primarily the upper limbs for postural risks. The REBA is a quick and systematic assessment that assesses the entire body for postural risks. Globally, MSD is the largest single cause of work-related illness; accounting for over 33% of all newly reported occupational illnesses in the general population and 77% in construction workers (Punnett and Wegman 2004). MSD is also the largest cause of sick absenteeism, days of work lost and disability. Very few studies are done in India related to the health issues and psycho social factors affecting constructions. By considering this, the research work focuses on the ergonomic risk factors that affect the construction workers and the relativeness of awkward postures to the health condition of workers.

6.2 Research Methodology

A construction site at NITK (National Institute of Technology Karnataka) was chosen for the research study and 25 male bar benders within an age range of 18–26 years were observed and studied for 30 days (February 2016). A semi-structured interview was conducted on the chosen bar benders. Bar benders were surveyed by means of a questionnaire as mentioned in Appendix 6.1 and the postural analysis is done by recording the working postures of the workers and by direct observation from the construction site. A comprehensive and semi-structured questionnaire is designed which is comprised of socio-demographic profile and job characteristics, discomforts of musculoskeletal systems. The participants were asked about complaints of the musculoskeletal system during the past six months (September 2015–February 2016) and during the last 30 days (February 2016), the perceived work-relatedness of the symptoms and the occupational tasks. Completing the questionnaires took approximately 30 min. The participants were interviewed and questionnaire is filled by the interviewer. The working postures of the workers are observed carefully and recorded for the duration of 30 days from the construction site at the working time. By direct observation the various tasks performed and the risk factors can be vaguely identified. The recorded postures are analyzed using the postural assessment tools REBA and RULA. The repetitive and difficult working postures of the bar benders are chosen for the postural assessment. With the help of the postural assessment tools, the risk level of the postures can be determined.

6.3 Results

6.3.1 Questionnaire Survey

The results from the questionnaire survey include the demographic data of the 25 bar benders who reported musculoskeletal discomforts in different body parts and the psychosocial factors that affect the workers. The demograpic variables, viz. age, work experience, height and weight the workers as collected are given in Table 6.1.

From the questionnaire survey conducted on 25 bar benders, it is observed that for the last 30 days discomforts in the shoulder reported is 48% (12) of the participants. The discomforts in upper back are 32% (8), low back is 16% (4), wrists 16% (4), hand 12% (3), knee 32% (8), ankles or feet 4% (1), and neck 4% (1).

Table 6.1 An overview of the collected demographic data

Demographic variables	Mean	Standard deviation	Range
Age (in years)	23	2.36	18–26
Work experience (in years)	3.16	1.54	1–6
Height of the workers (in cm)	164.88	6.34	154–175
Weight of the workers (in Kg)	55.04	5.78	45–68

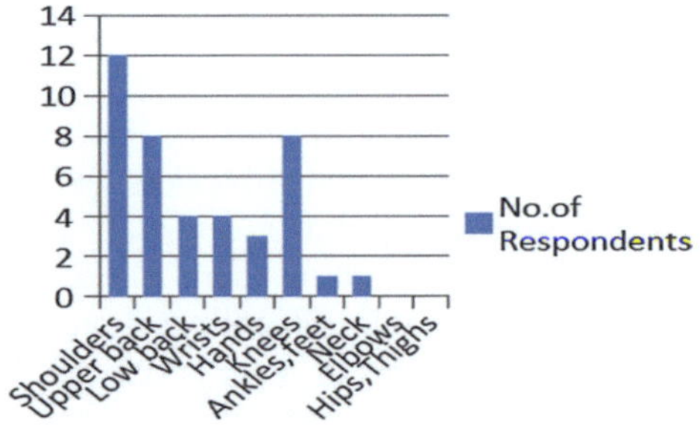

Fig. 6.1 No. of respondents versus affected body parts for the last month (February 2016)

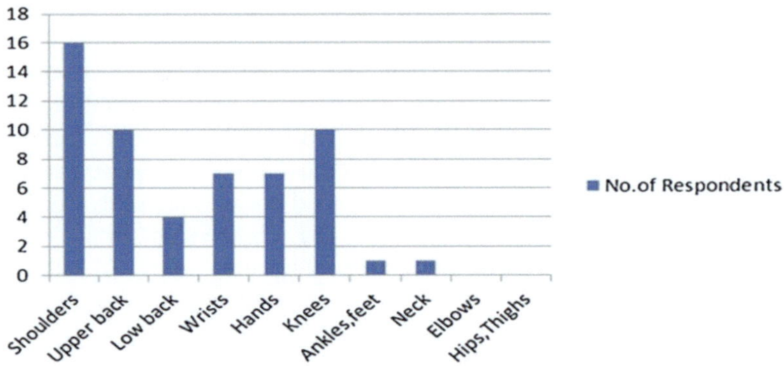

Fig. 6.2 No. of respondents versus affected body parts for the past six months (September 2015–February 2016)

There are no discomforts reported in the regions such as hips, thighs, and elbows. These discomforts considered for the last 30 days are lasted for at least 24 hours. The highest number of the workers is affected by the discomfort in the shoulder region. Manual weight lifting is shown as a main factor for the discomforts in shoulders. This type of works is less affected in the lower body parts except in knees.

It is also found that during the past 6 months, 64% (16) of participants got the pain in shoulder, in upper back it is 40% (10), in low back 16% (4), and in wrist it is 28% (7). About 40% (10) suffered from knee pain, 28% (7) suffered from pain in hand. Moreover, in ankles or feet and neck region, 4% (1) experienced discomforts as shown in Figs. 6.1 and 6.2.

Due to the musculoskeletal discomforts, a large percentage of workers are prevented from doing normal day-to-day works. Most of the workers are ignorant about the cause and effects of MSDs. 32% of the participants went to see a doctor because of MSDs and 40% of the sample took leave because of MSDs.

The questions regarding the psychological and social aspects were asked to the workers. The workers are uneducated or illiterate. Their unawareness and carelessness causes and increases the chances of affecting MSDs. The workers have a good team spirit and they prefer to work as groups.

6.3.2 *Postural Assessment by REBA and RULA*

The chosen 25 working postures of the bar benders are analyzed by the postural assessment tools REBA and RULA as shown below in Fig. 6.3. While doing the analysis by REBA, 60% of the postures are at medium risk, 32% of the postures are at high risk, 4% of the postures are at very high risk, and 4% of the postures are at low risk. While considering the analyzed postures, 96% of the postures come under medium, high, or very high risk and are recommended for investigation and to implement changes in the postures as shown in Tables 6.2 and 6.3 respectively.

While doing the analysis by REBA, 52% of the postures are at low risk, 36% of the postures are at medium risk, and 12% of the postures are at high risk. While considering the analyzed postures, 48% of the postures come under medim or high risk and are recommended for investigation and to implement changes in the postures. Table 6.4 represents the risk levels and various measures in order to minimize the risk of MSDs at various joints and segments of the workers.

6.4 Discussion

From the previous studies, it is proved beyond doubt that the awkward posture is one of the significant risk factors for causing MSDs. In the study among several workers with various skilled occupations, viz. masons, carpenters, brick-layers, etc., it is found that most of the bar bending workers suffer from MSDs in one or more body parts. A large number of workers experienced the difficulty in doing the work at site due to the pain in various joints and segments of the body. These problems that arise during works should be taken as a major concern while considering the improvement of the health and safety of the workers.

Twenty-five frequently repeated or difficult postures were analyzed by the postural assessment tools REBA and RULA. From the results of the postural assessment tools RULA and REBA, it is observed that 96% of the postures analyzed were at high or medium risk and recommended for changing the posture. It can be concluded that most of the working postures noted were critical and recommended for changing the posture.

Fig. 6.3 Various selected postures of bar benders

Fig. 6.3 (continued)

Table 6.2 Postural assessment by REBA

REBA score	MSD risk	No. of postures
1	Negligible risk	0
2–3	Low risk, change may be needed	1
4–7	Medium risk, further investigation, change soon	15
8–10	High risk, investigate and implement change	8
>11	Very high risk, implement change	11

Table 6.3 Postural assessment by RULA

RULA score	MSD risk	No. of postures
1–2	Negligible risk	0
3–4	Low risk, change may be needed	13

Table 6.4 Risk level and recommendations for the postures

Posture No.	REBA score	RULA score	Possible recommendations
1	5	4	Help the workers to modify the postures more comfortable
2	5	7	Change lifting method and make the workers to lift up to their capacity
3	4	3	Help the workers to modify the postures more comfortable
4	3	3	Same posture shall be continued
5	6	3	Help the workers to modify the postures more comfortable
6	4	3	Proper training should be given to the worker about lifting he materials
7	4	3	Help the workers to modify the postures more comfortable
8	4	4	Encourage the worker to choose comfortable posture
9	6	4	Ergonomic training should be given to choose the better posture
10	4	3	For working in proper heights, sufficient tools should be provided in the site
11	6	5	Avoid bending up to floor level by placing the materials and tools at comfortable heights
12	5	4	Help the workers to modify the postures more comfortable and provide good grips
13	5	5	Properly stack the materials at comfortable heights
14	10	7	Change lifting method and provide training
15	8	6	Ergonomic training should be given to identify the potential problems and to correct the working posture
16	9	4	Providing proper tools can avoid kneeling and if it is not possible to kneel pads should be used while kneeling
17	9	5	Use a rebar-tying tool with extended handle
18	10	5	Use a rebar-tying tool with extended handle
19	7	4	Ergonomic training should be given to choose the better posture
20	7	5	Help the workers to modify the postures more comfortable
21	9	5	Use a rebar-tying tool with extended handle
22	6	4	An upright posture should be adopted
23	5	3	Help worker to choose the comfortable posture
24	8	5	Better posture shall be adopted
25	11	7	Team lifting method or other alternative lifting methods should be used

The bending activities and weight lifting activities are found as the most critical activities which cause high risks. The workers with MSD should be in a closer look for their different occupational activities. For example, bar bending workers with MSD associated with the body parts like shoulders and low back should be taken into observation while carrying and lifting heavy materials.

From the compiled data, from the questionnaire survey and the REBA and RULA postural analysis tools, it can be concluded that due to the awkward postures adopted by the workers, the degree of probability for MSDs increases.

The most affected regions are shoulder, upper back, and knees. The use of human labor in moving weights, lack of ergonomic training regarding proper weight lifting and distribution all contributed to the involvement of these body parts.

6.5 Conclusions

The research shows that there exists a high degree of correlation between the ergonomic risk factors like awkward postures and work-related MSDs in the case of bar bending workers. The majority of the MSDs are reported as work-related by the bar bending workers.

Complains of the shoulder, knee, and upper back were most prevalent among the bar bending workers. Bending and weight lifting activities are considered as more critical activities which cause high risk.

The postural analysis using REBA and RULA methods categorized postures and represents the workers risk and indicates possible action to avoid or reduce the risk for MSDs. The RULA and REBA on the bar bending workers give strong evidence of the involvement of awkward postures in increasing the risk for MSDs.

6.6 Recommendations

The workers should work with an upright working posture and ergonomic principles should be used in tools and workplace design Safety and health awareness should be given to the workers. The workers should be encouraged by the supervisors to choose the comfortable working postures Repetitive postures, working in a position for the long term and exposing to excessive force should be limited and should keep under close observation Manual material handling should be reduced and new technologies and machinery can be introduced.

Appendix 6.1: Questionnaire Survey

Part 1: General information

1	Gender	
2	Age	
3	Job title	
4	How long have you been doing your present type of work?	
5	On average how many hours per shift do you work?	
6	How long is the time for rest breaks?	
7	Weight	
8	Height	
9	Do you have health problem	

Part 2: This part of survey is designed to determine the musculoskeletal symptoms among the workers

1. Have you ever been prevented from doing normal work (at home or away from home) because of MSD?

   Yes ☐ No ☐

2. Where was this pain or discomfort?

   ____

3. Did you see a doctor because of pain?

   Yes ☐ No ☐

4. Have you took leave because of this pain?

   Yes ☐ No ☐

5. What caused the pain?

   ____

6. What is the intensity of pain/discomfort?

   ____

7. How frequent the pain comes?

____

8. Do you feel any special kind of actions increases your pain? If yes, describe.

____

References

Armstrong TJ, Buckle P, Fine LJ, Hagberg M, Jonsson B, Kilbom A, Kuorinka IA, Silverstein BA, Sjogaard G, Viikari-Juntura ER (1993) A conceptual model for work-related neck and upper-limb musculoskeletal disorders. Scandinavian J Work Environ Health 73–84

Hignett S, McAtamney L (2000) Rapid entire body assessment (REBA). Appl Ergon 31(2): 201–205

Jaiswal N, Veerkumar V (2016) Work related musculoskeletal disorders among construction workers of India. Res J Fam Community Consum Sci 4(2):1–5

McAtamney L, Corlett EN (1993) RULA: a survey method for the investigation of work-related upper limb disorders. Appl Ergon 24(2):91–99

Parida R, Ray PK (2011) A comprehensive framework for physical evaluation of manual material handling tasks. Int J Manuf Technol Manage 24:153–166

Parida R, Ray PK (2012) Study and analysis of occupational risk factors for ergonomic design of construction work systems. Work 41(Suppl 1):3788–3794

Parida R, Ray PK (2015a) Factors influencing construction ergonomic performance in India. Procedia Manufact 3:6587–6592

Parida R, Ray PK (2015b) Biomechanical modelling of manual material handling tasks: a comprehensive review. Procedia Manufact 3:4598–4605

Punnett L, Wegman DH (2004) Work-related musculoskeletal disorders: the epidemiologic evidence and the debate. J Electromyogr Kinesiol 14(1):13–23

Ray PK, Parida R, Saha E (2015) Status survey of occupational risk factors of manual material handling task at a construction site in India. Procedia Manufact 3:6579–6586

Rwamamara R, Lagerkvist O, Olofsson T, Johansson B, Kaminskas KA (2010). Prevention of work-related musculoskeletal injuries in construction industry. In: Modern building materials, structures and techniques 2010 proceedings of the international conference in Vilnius, Lithuaniya, 2010, pp. 1293–1296

Sain MK, Meena ML (2016) Occupational health and ergonomic intervention in Indian small scale industries: a review. Int J Recent Adv Mech Engin 5(1):13–24

Chapter 7
Formulation of Mathematical Model to Investigate Magnitude of Musculoskeletal Pains Among Cotton Spinning Operators

Dhananjay Ikhar and Vishwas Deshpande

Abstract Hand-driven cotton spinning on traditional Charkha is the major small-scale entrepreneurship observed in Indian rural sector. The majority of women workers in these industries perform cotton spinning task adopting an uncomfortable posture. A cross-sectional study was conducted at two sites in central India region where 40 women operators were ergonomically evaluated to assess the effect of bad postures. A Dutch Musculoskeletal Questionnaire (DMQ) was used. It was observed that 50% operators were suffering from musculoskeletal pains in different body regions. Twisting, bending, and over-reaching were the resultant of poorly designed workstations. It is revealed that mathematical model must be developed to find micro influences of various independent parameters over dependent parameters like neck, upper back, shoulder, lower back, elbow, wrist, knee, hip, and ankle/feet. An exponential model is developed based on response data from DMQ and also validated using ANN simulation. For cotton spinning occupation, duration of occupation has a greater impact over MSDs in Hip region (5.13%) compared to MSDs in other body regions. Health status, in general, is found to be influential on neck (15.75%), hip (16.04%), wrist/hand (12.47%), and ankles/feet (12.35). It has significant influence over the development of MSDs in elbow (24.85%), neck (15.17%), and upper back (13.71%).

Keywords DMQ · Ergonomics evaluation · Musculoskeletal disorders · Mathematical modeling

D. Ikhar (✉)
Department of Mechanical Engineering, Datta Meghe Institute of Engineering, Technology and Research, Wardha, Maharashtra, India
e-mail: dhananjayikhar@gmail.com

V. Deshpande
Department of Industrial Engineering, Ramdeobaba College of Engineering and Management, Nagpur, Maharashtra, India

P.K. Ray and J. Maiti (eds.), *Ergonomic Design of Products and Worksystems—21st Century Perspectives of Asia*, Managing the Asian Century,
https://doi.org/10.1007/978-981-10-5457-0_7

7.1 Introduction

Musculoskeletal Disorders (MSDs) are a significant public health problem today, due to their high impact on disability, personal suffering, absence from work, and the direct and indirect costs to the healthcare system. A small-scale cotton spinning industry is identified in Wardha, Maharashtra, located in central India where operators are mostly women operators and 65% of them are suffering from one or more WMSDs. The majority of the work is performed on the cotton spinning wheel by the operator. The wheel is manually energized. It is a hand-powered device for spinning cotton yarn from pressure clamp. The spinning operation is performed in a squatting position in which operators, rotate spinning wheel sitting down on the hard and flat surface with folded knees without any backrest. Figure 7.1 shows the details of posture adopted.

The task of rotating the wheel for cotton spinning is repetitive and continuous throughout 8 h of working in a whole day. In this condition, the back is bent excessively and postures of different parts of body dramatically deviate from the neutral. Most of the experienced operators have left their jobs because of poor working conditions and musculoskeletal problems.

The overall purpose of this study is to present an overview of epidemiological evidence on the relationship between MSDs and workplace factors. Industries can use the data which is collected in this study to optimize the work systems to minimize the risk of injury development and to maximize the output quality.

The objective of the present research is to assess the magnitude of MSDs among cotton spinning operators. Based on the response data collected through DMQ, a mathematical model is formulated correlating responses in terms of independent variables. Indices obtained from these models were used to find influences of independent variables on the dependent variable.

Fig. 7.1 Working posture adopted in order to perform spinning operation

7.2 Statistical Analysis of DMQ Responses

Based on the responses of questionnaire, statistical analysis is carried out. The DMQ allows ergonomists and occupational health professional to measure work-related musculoskeletal risk factors and symptoms in worker populations in a quick and standardized way. The questionnaire seeks to obtain a simple representation of the relationship between work tasks and musculoskeletal symptoms. It is a standardized method relatively inexpensive and easy. It gives broad comprehensive overview of possible risk factors and morbidity (data on both exposure and effect). There is no need of any technical equipment to conduct this survey. Responses were collected from the workers themselves.

Total 40 female subjects were studied. The mean age was 38.5 years; mean height was 151 cms; and mean weight being 50.8 kg. The mean work experience was 8 years. On an average, they worked for 8 h every day. The lunch period is approximately about half an hour after the continuous work of 3 h (Table 7.1).

Perceived health status: Health status in general for study subjects has been assessed and it was found that 42.5% of the workers were in reasonably good health condition while 22.5% were observed in good health status. 27.5 and 7.5% workers were experiencing not bad and poor health conditions, respectively.

Perceived physical tiredness: Majority, i.e., 40% were experienced tiredness after the end of the day work. 12.5% were not tired and 35% were a bit tired after the work. 7.5% of the study subjects were experienced maximum tiredness at the end of the days' work.

Perceived pain status: Table 7.2 shows the related results in the percentage of worker experienced pain at different region of body in last 12 months. The body parts are also categorized further in left/right side and the pain values have been assessed for each of these separately. Majority of the study subjects i.e. 80% have experienced pain in right shoulder whereas left shoulder pain was experienced by 70%. Right wrist/hands and right knees (both 72.5%) was the second most painful site observed in the study population. Besides that left wrist/hands and left knees pain was observed in 57.5% of the workers. 65% of the study subject experienced pain in upper and lower back. Neck pain percentage was 52.5% for the duration of 12 months. Hips/thighs pain was contributed to 50% of the subjects for left and right sides both. Elbow and ankle/feet pain were observed in 42.5 and 37.5% of the workers, respectively, for both left and right sides of the body. Table 7.2 shows the percentage of study subject experiencing pain in last 12 months.

Table 7.1 Sociodemographic characteristics of study subjects

Variable	Mean	SD	Range
Age (years)	38.5	8.96	18–60
Height (cm)	151	11.45	145–165
Weight (kg)	50.8	8.9	45–60
Work experience (years)	8	4.69	1–18
Weekly working hours	48	1.35	45–50

Table 7.2 Physical complaints attributed to work by the study subjects

Variable	% pain experienced		
		Freq.	Percentage
Neck		21	52.5
Upper back		26	65
Lower back		26	65
Shoulders	Left	28	70
	Right	32	80
Elbows	Left	17	42.5
	Right	17	42.5
Wrist/hands	Left	23	57.5
	Right	29	72.5
Hips/thighs	Left	20	50
	Right	20	50
Knees	Left	23	57.5
	Right	29	72.5
Ankles/feet	Left	15	37.5
	Right	15	37.5

Table 7.3 Perceived workload and ergonomic environment

Type of load	Freq.	Percentage (n = 40)
Dynamic Load		
Moderate	15	37.5
Heavy	25	62.5
Repetitive Load		
Light	17	42.5
Moderate	20	50
Heavy	3	7.5
Static Load		
Light	2	5
Moderate	13	32.5
Heavy	25	62.5
Ergonomic Environment		
Good	17	42.5
Poor	23	57.5

Perceived workload and ergonomic environment: Majority of the workers (62.5%) perceived the dynamic load as heavy, half of the workers perceived the repetitive load as moderate while 62.5% workers perceived static load as heavy. Majority of them, i.e., 57.5% perceived the ergonomic environment as poor (Table 7.3).

7.3 Formulation of Mathematical Model

From the outcomes of statistical analysis, it is revealed to develop a mathematical model for the data collected through DMQ questionnaire. The data of the independent and dependent variables of the system has been gathered. A number of identical responses have been grouped together to reduce the number of independent variables, in turn, reducing the complexity of the model. In this case, there are 9 dependent and 28 independent variables. It is necessary to correlate quantitatively these independent and dependent variables involved in the experimentation. This correlation is nothing but a mathematical model as a design tool for such process. Details of these parameters are given in Appendix 7.1. Responses were collected on 5-point scale (Excellent, Good, Moderate, Poor, and Very Bad). Exponential trends forecast that the amount of growth will increase continuously. Thus models with a damped trend have been developed for longer-range forecasting. Some of the things where exponential growth is used to model include prevalence of MSDs among increased population in cotton spinning occupation. However, exponential model is robust; it gives reasonable precision. Exponential model is developed as follows.

For exponential relationship, Y_1 can be expressed as

$$
\begin{aligned}
Y_1 = k_1 \times x_1^{a_1} \times x_2^{b_1} \times x_3^{c_1} \times x_4^{d_1} \times x_5^{e_1} \times x_6^{f_1} \times x_7^{g_1} \times x_8^{h_1} \times x_9^{i_1} \times x_{10}^{j_1} \times x_{11}^{k_1} \\
\times x_{12}^{l_1} \times x_{13}^{m_1} \times x_{14}^{n_1} \times x_{15}^{0_1} \times x_{16}^{p_1} \times x_{17}^{q_1} \times x_{18}^{r_1} \times x_{19}^{s_1} \times x_{20}^{t_1} \\
\times x_{21}^{u_1} \times x_{22}^{v_1} \times x_{23}^{w_1} \times x_{24}^{x_1} \times x_{25}^{y_1} \times x_{26}^{z_1} \times x_{27}^{A_1} \times x_{28}^{B_1}
\end{aligned}
$$

After solving it for all 9 dependent variables, indices are formed.

From Fig. 7.2, it is found that neck pain develops in the operators majorly in the present occupation and it is because of sudden unexpected movements which

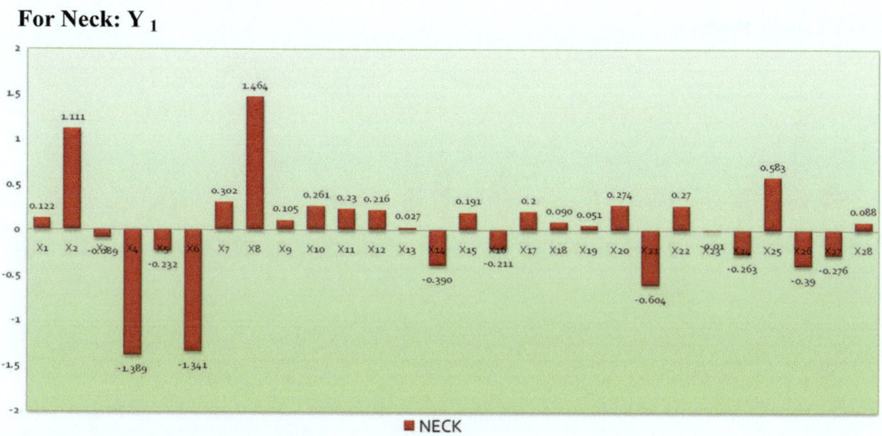

Fig. 7.2 Influences of independent variables on neck

indicate the highest index of 1.464. According to Binder (2007), in some industries, neck related disorders account for as much time off work as low back pain due to frequent unexpected movements. Neck pain causes severe disability in 5% of affected people leading to cervical spondylosis. Body mass index also contributes to MSDs in neck by index 1.114. Among the cotton spinning operators in this study, 15% are obese, 5% are thin, and rest are normal as per their BMI. After reexamining these operators, it was found that neck pain is very prominent among them.

From Fig. 7.3, it is found that cotton spinning operators are prone to lower back pain in the present occupation. It is developed due to poor health status (0.394), bent and twisted trunk experienced in working posture (0.587) which affects their body and daily routine (0.59). Again BMI (1.061) is found significantly contributing to the prevalence of lower back MSDs among operators. Viester et al. (2013) investigated the relationship between body mass index and musculoskeletal symptoms in working population. The association of BMI with all MSDs is found to be significant. He states that high BMI is associated with an increased 12-month prevalence of musculoskeletal symptoms. This association is significant for both overweight and obesity regarding overall musculoskeletal symptoms. Overweight is associated with upper and lower extremity symptoms.

It is found that cotton spinning operators are prone to shoulder pain due to sudden and unexpected movements (1.977) in the present occupation (Fig. 7.4). It is developed due to bend and twisted posture experienced (0.671). Low back pain (0.635) is also contributing to the development of pain in the shoulder. It seems that low back pain is radiating to shoulders due to twisted posture in operating machine in repeated manner. Ghosh et al. (2010) investigated that twisting, bending, and over-reaching are the resultant of the poorly designed workstation. These actions force the spine into a non-neutral position that increases the overall discomfort and pain particularly at the lower back, neck, and shoulders, which indicate that the charkha operators are be affected by work-related upper body MSDs. Metgud et al. (2008)

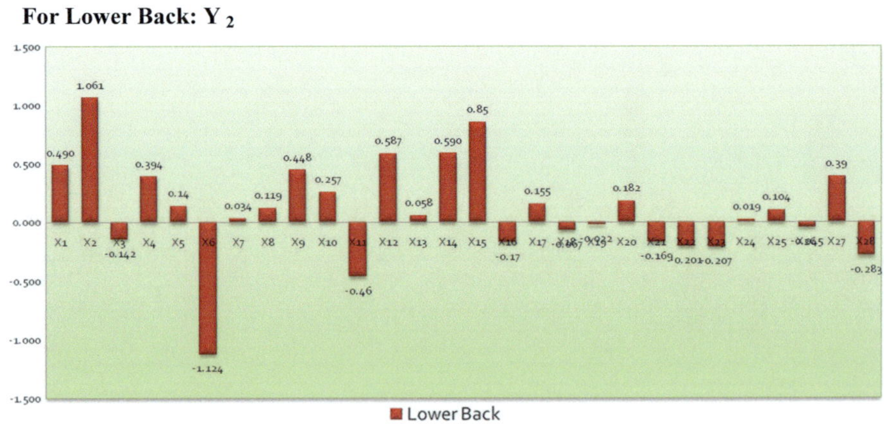

Fig. 7.3 Influences of independent variables on lower back

For Shoulder: Y_3

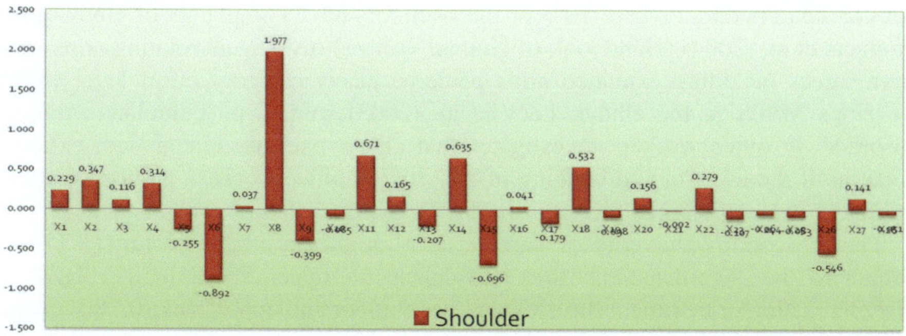

Fig. 7.4 Influences of independent variables on shoulder

For Elbow: Y_4

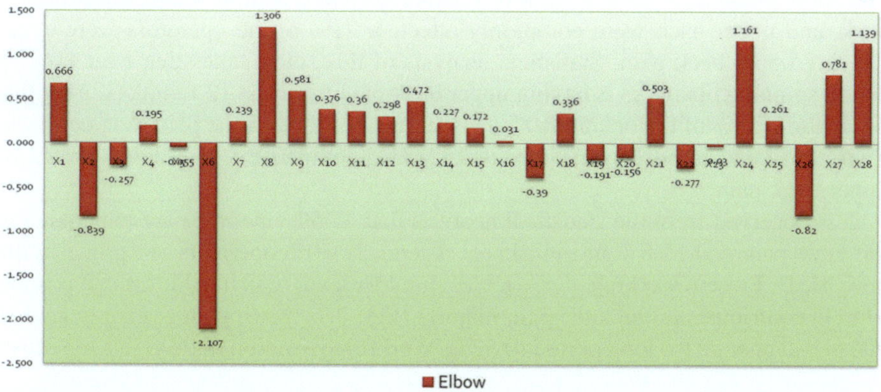

Fig. 7.5 Influences of independent variables on elbow

state that scapular muscles on the right side were involved in stabilizing the shoulder, which never went overhead. On the contrary, left shoulder was raised as high (>90°) in spinning action while pulling the thread. There was no pause since the wheel continued to spin the thread continuously unless workers opt to stop the work. These actions lead to MSDs in the shoulder.

All operators experienced elbow pain due to repetitive movements while operating charkha (Fig. 7.5). From statistical analysis of data, it is observed that in right-hand elbow the higher magnitude of pain was perceived than the left. Sudden and repetitive movements (1.306), the stress developed in neck and shoulder (1.161), and radiating pain from neck and shoulder are found to be very significant to the prevalence of MSDs in the elbow. Punnett et al. (1985) observe elbow pain in the workers working in garment factories. Mostly stitchers, under pressers were found to be more chronic to elbow pain because of overhanging of hand for long period and it also indicates the monotonous and repetitive work with hands may lead to the pain in elbow. Metgud et al. (2008) studied elbow flexion and extension for charkha operators where 70% of the operators have complaints of elbow pain. Rodgers et al. (2004) (Handbook of Human Factors) investigate rotating arms with high forces including extended arms perceive heavy category effort level which develops MSDs in the elbow. Lacy et al. (2003) studied on cumulative trauma disorders in office workers investigates that elbow pain associated with extreme rotation of the forearm and bending of the wrist in office workers. Cotton spinning operators also experience this posture during work

Present working above/under shoulder level or uncomfortable posture (1.348) is found to be significant in the prevalence of upper back pain (Fig. 7.6). Hospitalization or treatment for neck and/or shoulder and upper back (0.708) is also observed from the indices obtained. Correlation of lower back pain and upper back pain can be developed through the obtained values. The radiating pain of lower back to upper back is observed which contributes to the disturbance of daily routine of the operators. Choobineh et al. (2004) investigated MSDs in carpet mending operation where he founds for 8 h of continuous and repetitive work, knees, lower back, and upper back were commonly affected. 21% of the operators were experiencing upper back pain. Statistical analysis of this study states that over 80% of cotton spinning operators is having upper back pain from last 12 months during this occupation. 19% of the operators also visited the hospital for the pain in upper back. Twist and bend in neck or posture (0.352) also contribute to the development of upper back pain.

It is observed from the statistical analysis that 47.5% operators are experiencing left knee pain and 52.5% in right knee. Averagely 50% operators are prone to the knee MSD. Present working above under shoulder level or in uncomfortable posture (0.811) contributes to the knee pain majorly (Fig. 7.7). Knee pain is mainly caused due to bad posture for long period (0.324). Workforce exertion (0.264) in a repetitive

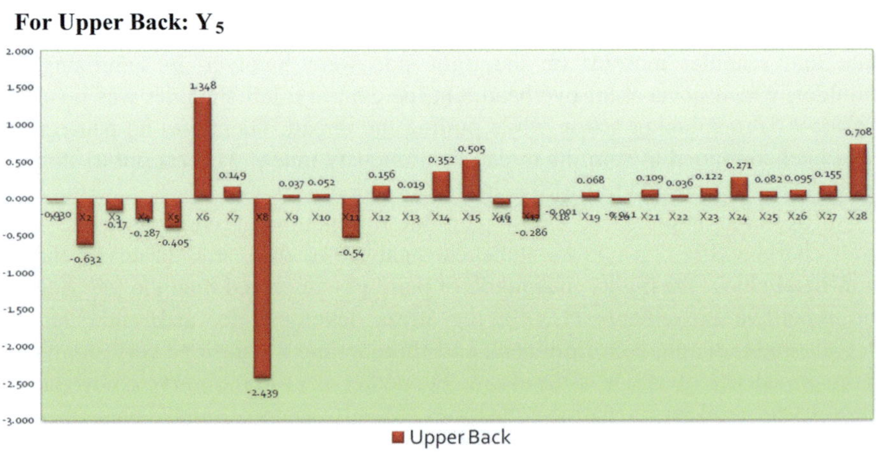

Fig. 7.6 Influences of independent variables on upper back

For Knee: Y 6

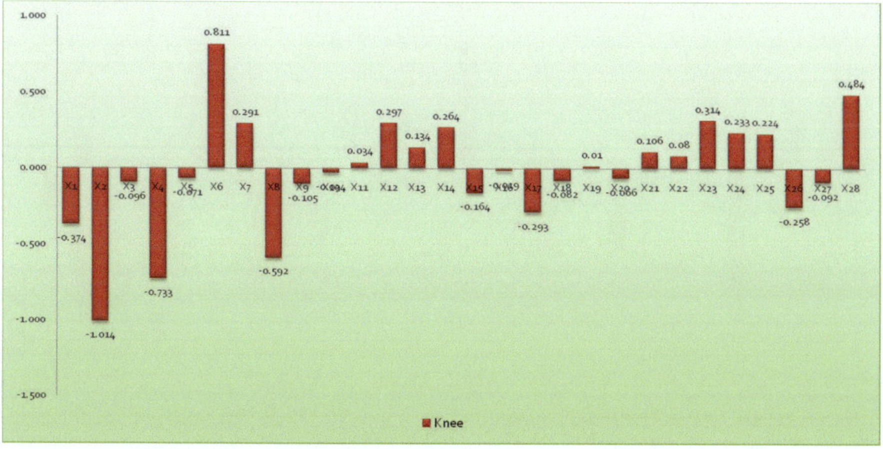

Fig. 7.7 Influences of independent variables on knee

manner leads to MSD in the knee. Choobineh et al. (2004) state that similar kind of seating posture is adopted in carpet mending operation and around 30% of the workers experiencing knee pain in occupation. Nag et al. (2008) explored the applicability of stabilometry in evaluating the floor-seated postures, which women adopt in domestic and cottage-based occupational spheres in certain parts of the world. The analysis showed that the sitting postures with extended legs caused the body to move largely in the AP direction, with a relatively greater velocity of displacement, indicating instability in the postures leading to MSDs in the knee.

Wrist and hand pain is observed in study population due to bent and twisted wrist (0.767) experienced by the operators while rotating the handle of the machine. Past work squatting and kneeling in uncomfortable posture (0.517) is observed in 13% of workers from their responses. BMI (0.347), low back pain severity (0.409) and bent and twisted trunk (0.44) are found to be contributing to wrist/ hand pain among operators (Fig. 7.8). Awkward postures of wrists and fingers during spinning operation are mainly due to poor hand tool design and the repetitive nature of the tasks. Wrist extension/flexion while cutting rotating handle of the spinning machine are commonly observed in cotton spinning. Choobineh et al. (2004) also observed the MSD in wrist due to similar kind of operation in carpet weaving. Banerjee et al. (2003) studied on upper extremity CTD in different unorganized sectors in West Bengal. It is observed that a highly repetitive activity with non-neutral wrist posture, flexion, extension, forceful exertion with heavy tools used for a long period is a major cause of MSDs in wrist or hand. Statistical analysis shows that 42.5% operators have right wrist pain while 27.5% experienced left-hand wrist pain while handling tools.

It is found that low back pain (1.072) is majorly contributing to the prevalence of hip pain among operators (Fig. 7.9). Statistical analysis shows that all operators are

For Wrist/Hand: Y_7

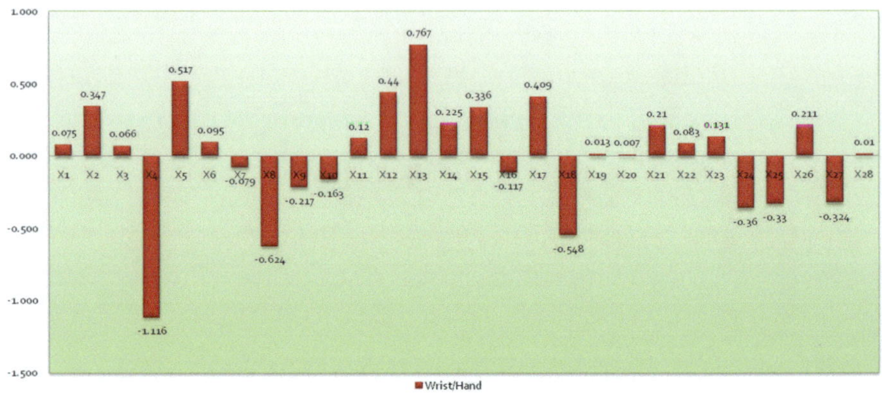

Fig. 7.8 Influences of independent variables on wrist/hand

For Hip: Y_8

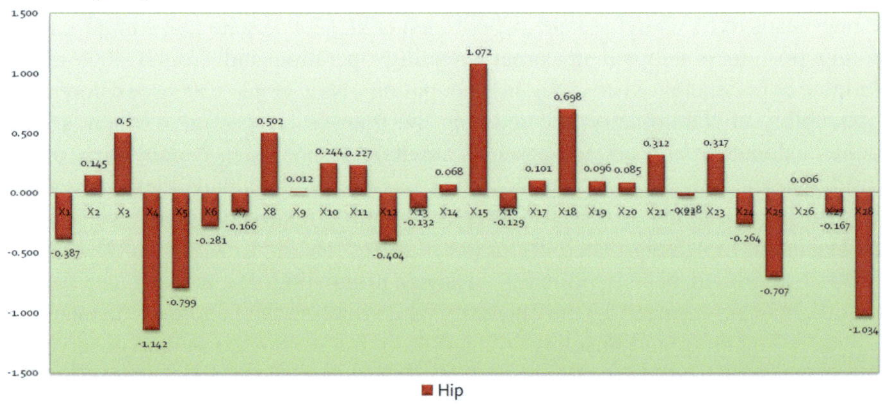

Fig. 7.9 Influences of independent variables on hip

experiencing hip pain due to sitting on the hard and flat ground surface with whole body weight and dynamic load conditions. Duration of occupation (0.5) in cotton spinning is also affecting this body region. Operators working for more than 3 years in this occupation (67%) are more prevalent to MSD in hips. Sudden and unexpected movements (0.502) along with the repetitive motion of hands are producing hip pain. Murrel (1965) states that basically when one sits down on a hard surface without any backrest, the ischial tuberosities (inferior protuberances of the pelvic bones) acts as fulcra around which the pelvic girdle rotates under the weight of upper body which generates pain in lower back and the same is radiated to buttocks.

It is observed from the indices that pain in feet is mainly due to sudden and unexpected repetitive movements (2.764) in cotton spinning operation (Fig. 7.10).

For Ankle/Feet: Y_9

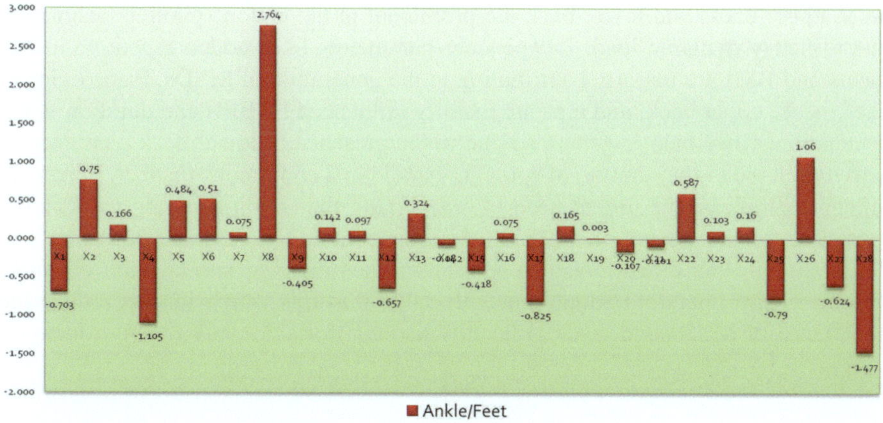

Fig. 7.10 Influences of independent variables on ankle/feet

65% of the operators are experiencing MSD in ankle/feet due to prolonged sitting posture at the workstation. It is also found that neck and/or shoulder pain severity (1.06) has a contribution to the development of ankle pain. Averagely 38% of the operators reported that neck and shoulder pain is getting worse day-by-day. BMI (0.75) has a significant correlation with MSD in Ankle/Feet.

The results obtained through the research work are found significant and are very well supported by various literature regarding the same. The outcomes of this research have been validated with the findings of similar types of workstation world over.

7.4 Conclusion

The research intends to draw the attention towards the risks that the hand-driven cotton spinning operators in India are facing due to bad design of workstation. The research has proved that the health of the operators is being significantly affected and there exists a very high degree of chance that these operators would generate MSDs in various body regions. Obtained results indicate areas of improvement in the posture of spinning operators and the workstation design. Through this research, one is able to identify the most hazardous working posture that needs immediate investigation and improvement. The majority of the subjects in this study had shoulder, back, and wrist pain which is due to repetitive nature of the job and the poor design of spinning workstation. Musculoskeletal complaints are prominent in our study because the female subjects study participants adopt prolonged sitting posture without any backrest and maintain this position for at least 8 h a day. As there are residual pain and fatigue, the productivity slows down. From the performance analysis of the mathematical model, it is investigated that MSDs in neck, upper back, and lower back are prominent in the cotton spinning occupation due to heavy dynamic load. Independent parameters like sudden repetitive movements and BMI are majorly contributing to the generation of MSDs. Pains in lower back, neck, upper back, and hips are majorly influenced by BMI and duration of the occupation. Knee pain is generated due to the presence of lower back pain and the activity related to the rotation of spinning handle in a continuous manner. Perceived influences indicate the urgent need of redesigning the cotton spinning workstation to reduce the prevalence of MSDs in this occupation. The research outcomes will definitely help the practitioners to take urgent actions over similar kind of workstations where operators adopts uncomfortable postures which in turn reduce the prevalence of MSDs and leads to healthy and comfortable working environment.

Appendix 7.1

Sr. No.	Name	Abb.
Independent variables		
1	Age	X1
2	BMI	X2
3	Duration of occupation	X3
4	Health status	X4
5	Past work squatting kneeling uncomfortable posture	X5
6	Present work above under shoulder level or uncomfortable posture	X6
7	Present work force exertion	X7
8	Present work sudden unexpected movements	X8
9	Repetitive movements in hand, trunk and head (neck)	X9
10	Bent and twisted neck	X10
11	Bent and twisted posture	X11
12	Bent and twisted trunk	X12
13	Bent and twisted wrist	X13
14	Low back pain affecting body and routine	X14
15	Low back pain associated with present work	X15
16	Low back pain cause bad posture long period	X16
17	Low back pain severity	X17
18	Low back pain start during present work	X18
19	Low back pain trouble when moving load >5 kg	X19
20	Radiating low back pain to left and/or right knee in past 12 months	X20
21	Ever had hospitalization or treatment for lower back pain	X21
22	Neck and/or shoulder associated with present work	X22
23	Neck and/or shoulder cause bad posture long period	X23
24	Neck and/or shoulder cause stress	X24

(continued)

(continued)

Sr. No.	Name	Abb.
Independent variables		
25	Neck and/or shoulder pain affecting body and routine	X25
26	Neck and/or shoulder severity	X26
27	Neck and/or shoulder when working in bad posture and tools	X27
28	Ever had hospitalizations or treatment for neck and/or shoulder	X28
Dependent Variables		
1	Neck	Y1
2	Lower back	Y2
3	Shoulder	Y3
4	Elbow	Y4
5	Upper neck	Y5
6	Knee	Y6
7	Wrist/hand	Y7
8	Hip	Y8
9	Ankle feet	Y9

References

Banerjee P, Gangopadhyay S (2003) A study on the prevalence of upper extremity repetitive strain injuries among the handloom weavers of West Bengal. J Hum Ergon (Tokyo) 32(1):17–22

Binder AI et al. (2007) Cervical spondylosis and neck pain. BMJ 334(7592):527–531

Choobineh A et al (2004) Ergonomic intervention in carpet mending operation. Applied Ergonomics 35:493–496

Ghosh T, Das B, Gangopadhyay S (2010) Work-related musculoskeletal disorder: an occupational disorder of the goldsmiths in India. Int J Community Med 35(2):321–325

Lacy CR et al (2003) Cumulative trauma disorders in office, public employees occupational safety and health program, VDT Guidelines. N.J. Department of Health and Senior Services

Metgud DC, Khatri S, Mokashi MG, Saha PN (2008) An ergonomic study of women workers in a woollen textile factory for identification of health-related problems. Indian J Occup Environ Med 2(1):14–19

Murrel KFH (1965) Ergonomics man in his working environment. 5th edn. Chapman and Hall Ltd

Nag PK, Vyas H, Nag a, Pal S (2008) Applying stabilometry in characterising floor sitting modes of women. Int J Ind Ergon 38(11):984–991

Punnett L, Robins JM, Wegman DH, Keyserling WM (1985) Soft tissue disorders in the upper limbs of female garment workers. Scand I Work Environ Health 11:417–425

Rodgers SH et al (2004) Muscle fatigue assessment: functional job analysis technique. Handb Hum Factors Ergon Methods 12–11

Viester L, Verhagen EALM, Oude Hengel KM, Koppes LLJ, Van der Beek AJ, Bongers PM (2013) The relation between body mass index and musculoskeletal symptoms in the working population. BMC Musculoskeletal Disorders 14:238

Chapter 8
Ergonomic Study on Work Postures of Sewing Machine Operators in Government Industry: A Case in Lean Environment Garment Industry

T. Sakthi Nagaraj and R. Jeyapaul

Abstract Several studies globally have highlighted musculoskeletal risk factors associated with the garment making jobs because of highly repetitive work in awkward work postures. The purpose of the research is to identify possible ergonomic deficiencies in the standing modular workstation of sewing machine operators in a lean environment garment industry. The discomfort analysis by Cornell Musculoskeletal Discomfort Questionnaire (CMDQ) and also the Rapid Entire Body Assessment (REBA) was carried out among a sample of experienced female operators in a garment industry located in Sri Lanka. From this study, compared to other parts, foot, knee, lower leg, neck, shoulders, thigh, and lower back are experiencing ache highly. It is concluded that the shoulder, upper back, upper arm, neck, forearm, and wrist are experiencing aches due to the machine is not ergonomically fit with the operator's body dimensions, and ache in the leg and lower back due to prolonged standing and working. The ergonomic solutions like fixing the proper height of sewing machine, inclined work table, providing anti-fatigue mats, job rotation, stretching exercises, and short-break approach are suggested for minimizing the absenteeism and turnover due to bodyache and to increase the efficiency and quality of the product and process.

Keywords Ergonomics · Modular sewing production · Lean environment · Cornell musculoskeletal discomfort questionnaire · Rapid entire body assessment

---

T. Sakthi Nagaraj (✉) · R. Jeyapaul
Department of Production Engineering, National Institute of Technology, Trichy 620015, Tamil Nadu, India
e-mail: sakthiamie@gmail.com

P.K. Ray and J. Maiti (eds.), *Ergonomic Design of Products and Worksystems—21st Century Perspectives of Asia*, Managing the Asian Century,
https://doi.org/10.1007/978-981-10-5457-0_8

8.1 Introduction

8.1.1 Musculoskeletal Disorders

Musculoskeletal disorders (MSDs) are a foremost health problem among the workers in both developed and developing countries (Widanarko et al. 2014). Work-related MSD issues have always been an area of examination and enhancement especially for industries involving work-intensive operations and techniques (Spasojević Brkić et al. 2015). MSDs are more common among women than among men (Dahlberg et al. 2004). Work-related MSDs have been recognized as a significant problem within the European Union in terms of individual ill health and economic factors such as the impact on production and revenues (Buckle and Jason Devereux 2002) and another part of the world also (Oh et al. 2011). Prevention of MSDs is most important factors that can have a major impact on productivity improvement and on the upgradation of health and safety at workplace (Kogi et al. 2003). Work-related musculoskeletal disorders (WMSDs) are a major anxiety in the industry which can also compromise competitiveness due to costs related to worker compensation, labor turnover, absenteeism, poor quality, and reduced productivity (Andersson 1992). There are various factors which can cause musculoskeletal disorders, such as low temperature and humidity. Improper work posture and repeated tasks in such an environment may harm the musculoskeletal system (Oksa et al. 2002; Piedrahíta et al. 2004). And also the fatigue failure plays an important role in the etiology of MSDs (Gallagher and Schall 2016). Finally some psychosocial and organizational factors and also some individual factors were associated with prevalence of MSDs (Piranveyseh et al. 2016).

8.1.2 Sewing Machine Operators (SMO)

The possible ergonomics deficiencies have been identified and described in the workstation of sewing machine operators in a textile industry in Botswana as well as their insight of workload and discomfort and also with specific reference to sewing machine operators, this is an occupational group that may experience a high occurrence of MSDs (Sealetsa and Thatcher 2011). The poor working postures of the workers have to be maintained during the whole working period as well as to repetitive hand and arm movements. Their posture is constrained by the task and the design of the sewing machine and table have a considerable influence on the posture adopted (Li et al. 1995). The sewing machine operators use both hands to handle workpieces, or feel objects, and tools; control by leg; sit and repeat the same motions for working periods of time. The sewing machine operation includes tasks that involve operating power sewing machines to sew, alter or repair wearing apparel, linens, blankets, and other fabric. This includes the operation of automatic sewing machines when the operator must know how to thread the machine, wind bobbins, adjust tension, and oil parts (Kaergaard and Andersen 2000). For this reason, a high occurrence of musculoskeletal symptoms of neck, shoulder, back, hand/fingers, and lower extremities have been found in studies of female SMOs (Kaergaard and Andersen 2000; Brisson et al. 1989; Kilroy and Dockrell 2000). Typically, sewing machinists perform piecework activities, assembling component parts of a final garment or product. The fragmentation of work in this way means that the task is often highly repetitive, potentially dangerous and complex task involving the coordination of both hands, usually performed in a seated working posture for a long period of time. The operation usually requires the operator to lean forward, with a forward inclined posture of both head and trunk, to focus their attention and have better visual control of the task. In many ways, given the fixed nature of the sewing machine, the machinists have to adapt their posture and viewing angle so that they can conduct this form of 'precision work' and this can lead to the development of MSDs, particularly in the back, neck, and upper limbs.

8.1.3 Modular Sewing Production

The decision to move straight-line assembly systems to U-shaped assembly lines systems constitutes a major layout design change and huge investment for assembly operations. U-shaped assembly systems offer several benefits over traditional straight-line layouts including an improvement in labor productivity, quality of the product, traceability, reduced WIP and reduced lead time and changeover time. The workers can move between the two legs of the U-line to perform combinations of tasks (Aase et al. 2004). The operators are standing and operating the machines to convenient to move from machine to machine and also it reduces the long-term static posture of the workers. The MSDs are the major health problem among the sewing machine operators working in the textile industries. Several ergonomic researches have been done for SMOs, who are all working in sitting position. Most of the aches are recorded in the upper body, hand, and arm and very less on the lower body (Oh et al. 2011; Sealetsa and Thatcher 2011; Li et al. 1995; Dianat et al. 2015). In the MSP, the SMOs are standing, operating and controlling the machine with the foot. There is no ergonomics studies have reported on standing modular SMOs.

Prolonged standing at workstations can cause muscle fatigue and mental stress. Furthermore, an insufficient rest period during the standing time coupled with improper footwear can lead to discomfort and fatigue in the lower extremities, causing occupational injuries in the long term. When workers work in a prolonged standing posture, static contraction occurs particularly in their back and legs, resulting in impaired functioning of calf muscles (Krijenen et al. 1997). Prolonged standing transfers the weight of upper body parts to lower parts and results in lower back pain. 83% of industrial workers in the USA experience foot or lower leg pain and discomfort associated with prolonged standing (Zander et al. 2004).

8.1.4 Rationale

Witd an overview of the modular sewing production in textile industries and the potential of this work to lead to MSDs provided, research was conducted to investigate the following factors:

- To determine MSD prevalence among the SMOs working in the lean environment by using Cornell Musculoskeletal Discomfort Questionnaire.
- To examine the level of ergonomic workplace risk factors by using the rapid entire body assessment.
- To develop the productivity by eliminating the ergonomic risks in lean environment.

8.2 Materials and Methods

8.2.1 The Workplace

This ergonomic study was conducted in a textile industry that manufactures children wear, and which is located in Sri Lanka. The main customers for the company are USA, Europe, and UK and suppliers from India, China, and Sri Lanka. To meet the market demand and agility, the industries have to adopt lean manufacturing system. Lean sets out a methodology for being highly responsive to customers' demands whilst constantly challenging costs and wastes throughout supply networks (Bhamu et al. 2014; Shah and Ward 2007). So, it would appear that Lean can be applicable to all sizes of enterprise in their endeavors to become more competitive to sustain, and possibly enhance, their position in the modern marketplace. The industry started to follow the standing MSP system, which consists of U-shaped layout, Multi-skilled operators, standing operations machines and single piece flow. The totals of 896 female SMOs are working in both shifts and shift duration is 8 h which includes half an hour total break time. The sewing floor consists of 32 modules and 14 SMOs for each. Each SMO sews the ready-to-sew cut panels coming from the cutting department. The full garment packed at the end of the layout and send to quality audit than Finished Goods Storage (FGS) through metal detectors. There were no ergonomic programs being conducted at all.

8.2.2 Participants

The study samples are paid based on their years of employment, an average of 100 USD per month. Based on their performance, incentive money also paid with their

Fig. 8.1 Standing modular sewing machine operators

monthly salary. The participation was on a voluntary basis and the operators were under no compulsion to complete the study. The operators were not paid for their participation. The primary task is the 'sewing' and the secondary task is trims and accessories handling apart from these tasks thread changing and cleaning tasks have to carry out by these SMOs. The sewing process carried out by flexed left and right side, standing by the left leg and operating the machine pedal with right leg throughout the shift (Fig. 8.1). The SMOs standing on the floor, leans forward at an angle of 25°–30° and head around 35°, and moves her hands backward and forward and to the left and right (Average 150 times per hour) to perform the sewing process. Most of the operators are not using footwear on the floor while sewing.

8.2.3 Research Design

The proposed research methodology is shown in Fig. 8.2. Initially the data collection about demographic and job characteristics among the sample SMOs (n = 557) working in the case industry. The chi-square test was conducted to identify the significant factors for the discomfort among the demographic and job characteristic factors. The operators (n-336), who were all felt discomfort demographic and job characteristic survey, were considered to further discomfort survey CMDQ and posture analysis REBA and who did not feel discomfort were not considered for further study (n-221). The CMDQ and REBA validated by test–retest and reliability test, respectively. These study outcomes have analyzed and discussed further.

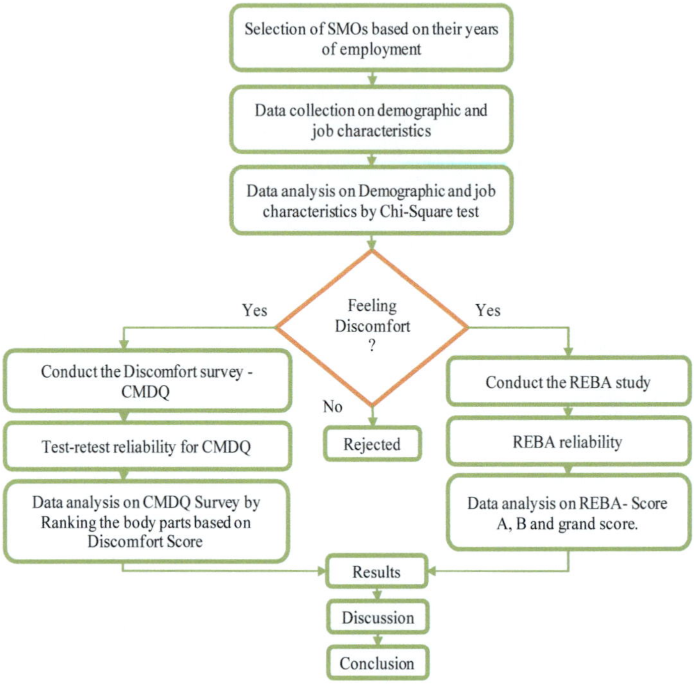

Fig. 8.2 Research methodology

8.2.4 Tools and Data Collection

In this study, the data were obtained with a questionnaire and by direct observation, done by the trained management trainees. The management trainees are undergraduate fresher working in the case company and properly trained about these survey and tools. The data collection process consists of three parts; one is about demographic and job characteristics, the second one is about discomfort survey, and the final one is posture analysis. The 557 SMOs from both shift (Out of 896 SMOs), have selected for demographic and job characteristics study based on their years of employment in the case organization.

8.2.4.1 Demographic and Job Characteristics

In the first part the questionnaires, information was collected about age, weight, height, Body Mass Index (BMI), education, marital status, daily walking distance, SMO experience, job satisfaction, job stress, and finally the discomfort of the SMOs. The information has collected by interviews with the SMOs. The age (in years) considered as a factor and BMI, based on the height and body weight, the

BMI (kg/m^2) was calculated. Marital status included because of married SMOs have to contribute their household works and for daily walking distance included because of a lot of SMOs have to walk (2–3 km) to reach the transport boarding place or to the company. The SMOs experience is considered only on standing operation in MSP. The discomfort of the SMOs was answered 'Yes or No' and job satisfaction and job stress, in "low, medium & high" scale.

8.2.4.2 Cornell Musculoskeletal Discomfort Questionnaire (CMDQ)

The list of SMOs, all who felt discomfort in the previous study, have been considered for further discomfort survey and REBA analysis. The CMDQ consists of three sectional parts and each consists of a different scoring part (*i*), shows in Fig. 8.3. Respondents indicated the frequency of discomfort on an ordinal scale (*R*) from 0 (Never) to 4 (Several times everyday) and severity of discomfort from 1 (slightly uncomfortable) to 3 (very uncomfortable). The level at which the discomfort interfered with work was scored from 0 (Not at all) to 2 (substantial interference). Total discomfort score was calculated by using the following formula (Jansen et al. 2012) (2). Here, the interference score starts from 1 (Not at all) to 3

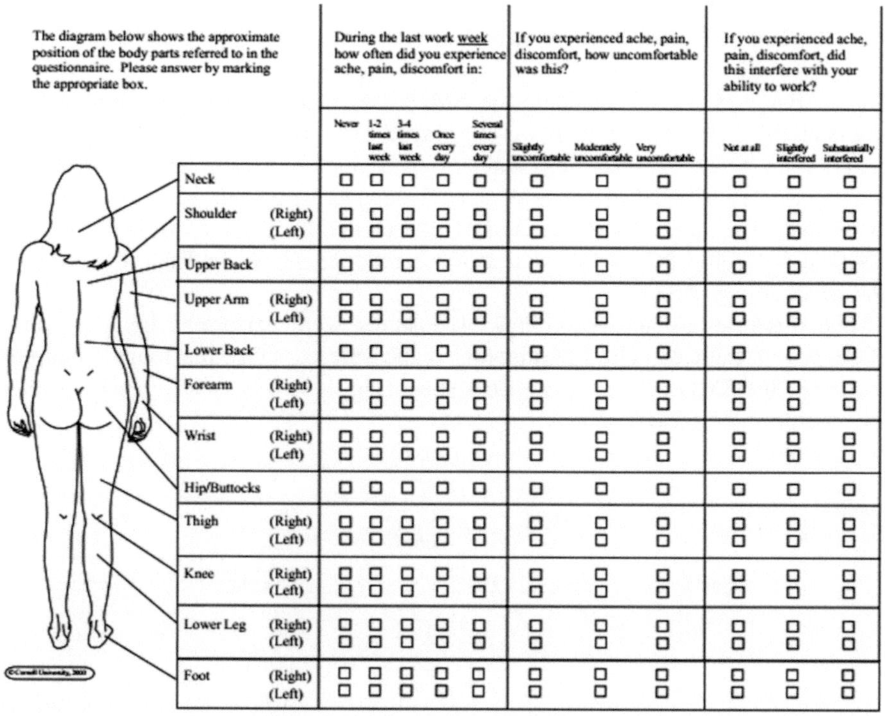

Fig. 8.3 Cornell musculoskeletal discomfort questionnaire (CMDQ), female version (http://ergo.human.cornell.edu/ahmsquest.html)

(substantial interference) proposed. The frequency (*F*), severity (*S*) and interference (*I*) score of a particular body part were calculated (1) as the addition of the number of SMOs multiplied by corresponding ordinal scale score.

$$F(or)S(or)I = \sum_{i=1}^{r} n_i R_i \tag{8.1}$$

where n_i is the number of operators (n) in *i*th scoring part and R_i is the corresponding score in ordinal scale.

$$Total\ Discomfort\ score = Frequency \times Severity \times Interference \tag{8.2}$$

The total discomfort score of different body parts will be calculated from the above Eq. (8.2). Test–retest reliability for CMDQ at a one-month interval found a 3% difference in responses for upper body parts and a 7% difference for lower body parts. This study was done among 30 SMOs. The questionnaires were filled in at the workplace. This CMDQ survey was conducted among the SMOs who were answered 'Yes' for discomfort question in previous (demographic and job) characteristics survey. The 336 SMOs have answered that feeling discomfort.

8.2.4.3 Rapid Entire Body Assessment (REBA)

The REBA was used to evaluate the SMOs full body working postures at their workstations. The body parts were grouped into two categories. "Group A" included the trunk, neck and legs and "group B" included upper arms, lower arms, and wrists. Each body part was evaluated depending on the angle of parts and load/force. The scores were calculated to get the final score, which was then assessed according to the proposed action categories shown in Table 8.1 (Hignett and McAtamney 2000).

The observations and recordings of postures were performed by ten trained management trainees. The investigators made sure to study the longest adopted posture of the SMOs. A separate REBA sheet (Fig. 8.4) was used for each operator for recording the REBA scores.

Table 8.1 REBA—Action Categories

Action level	REBA score	Risk level	Action
0	1	Negligible	None necessary
1	2–3	Low	May be necessary
2	4–7	Medium	Necessary
3	8–10	High	Necessary soon
4	11–15	Very high	Necessary NOW

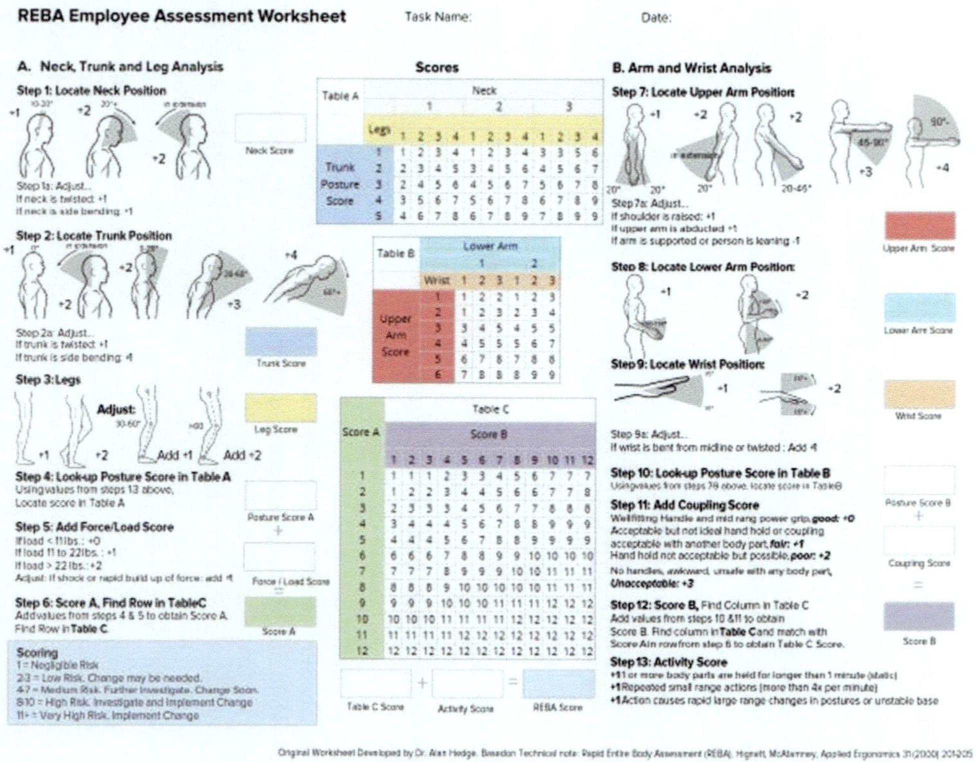

Fig. 8.4 REBA Work sheet ('REBA-Worksheet-v-2.0.pdf', 15 July 2015)

8.2.4.4 REBA Reliability

To establish how much homogeneity or consistency in the REBA score, which collected by the management trainees. The reliability of the REBA scores, observed by each management trainees, using Cronbach's Alpha was found to be 'good' ($0.9 > \alpha \geq 0.8$) (with Cronbach's Alpha 0.86 for REBA scores).

8.2.5 Data Analysis

Statistical analysis was performed with Minitab 16. The analytical statistics were carried out using chi-square test (Öztürk and Esin 2011) and used to assess relationships between demographic (Age, BMI, education, marital status, daily walking distance,) with job-related characteristics (SMO experience, job satisfaction & job stress) and the discomfort. The CMDQ survey also analyzed in terms of individual body parts of total SMOs and frequency, severity and work interference of case industry. The REBA score was analyzed in terms of percentage of operators in risk level. For all statics test the *P* values below 0.05 were considered statistically significant (Dianat et al. 2015).

8.3 Results

8.3.1 Demographic and Job Characteristics

Table 8.2 presented the demographic and job characteristics details of the study participants. The study sample comprised of 557 SMOs. The ages of the SMOs were ranges from 18 to 36 (Mean-25.9, Standard Deviation (SD)-4.4) and the majority of them were in the range of 21–25(37.2%), next one in 26–30(37%) and, above 30 and below 21-year-old was very minimum (<14%). The BMI of the SMOs were varied from 15.6 to 28.9 kg/m^2 (Mean-21.6, SD-2.7), the 58% of SMOs had healthy body, 25.5% overweight body, underweight (15%) and very less in obese (1.4%). The 57.8% of the operators were primary school level, 25.3% were secondary school level and higher secondary very less (16.7%). The 79.7% of the workers were married (Unmarried-20.3%) and operators daily walking distance to reach boarding place or company below 1 km (56.9%), 1–2 km (19.4%) and 2–3 km (23.7%). The experience only in MSP was considered in the study as a factor and ranges from 0.5 to 3 years. Most of the operators have below one year (49.9%) and 33.4% has 1–2 years and very fewer operators (16.7%) has 2–3 years. The most of the SMOs were satisfied with their job in medium level (50.8%) and fewer in high level (27.3%) and low level (21.9%). The stress due to job demand among the SMOs was medium (36.8%), low level (36.1%) and less in high level (27.1%).

Table 8.2 Demographic and job characteristics and their relationship to discomfort (n: 557)

Discomfort	n	Yes	No	Statistics	df	Significant	
		60.3% (n = 336)	39.7% (n = 221)				
		%	%				
Age							
Less than 21	69	60.9	39.1	2.35	3	$p > 0.05$	Not Significant
Between 21–25	206	56.3	43.7				
Between 26–30	207	62.8	37.2				
Above 30	75	64.0	36.0				
BMI							
Underweight (<=18.4)	84	53.6	46.4	8.68	3	$p < 0.05$	Significant
Healthy (18.5–22.9)	323	57.6	42.4				
Overweight (23–27.4)	142	70.4	29.6				
Obese (>27.5)	8	62.5	37.5				
Education							
Primary School	322	56.2	43.8	5.73	2	$p > 0.05$	Not Significant
Secondary School	93	65.6	34.4				
Higher Secondary School	141	66.7	33.3				
Marital status							
Married	444	60.6	39.4	0.063	1	$p > 0.05$	Not Significant
Unmarried	113	59.3	77.6				
Walking Distance to Boarding place							
Below 1 km	318	54.7	42.5	11.34	2	$p < 0.05$	Significant
1–2 km	118	53.4	41.5				
2–3 km	141	69.5	26.2				
Experience in MSP							
Below 1 Year	278	57.6	42.4	3.04	2	$p > 0.05$	Not Significant
1–2 Years	186	60.8	39.2				
2–3 Years	93	67.7	32.3				
Job Satisfaction							
Low	122	72.1	27.9	9.23	2	$p < 0.05$	Significant
Medium	283	57.6	42.4				
High	152	55.9	44.1				
Job stress							
Low	205	40.0	60.0	56.02	2	$p > 0.001$	Significant
Medium	201	72.6	27.4				
High	151	71.5	28.5				

8.3.2 CMDQ Discomfort Survey

Table 8.3 shows the discomfort, severity, and interference of SMOs with respect to different body parts and in terms of percentage of operators (n = 336). Among 336 SMOs, 85.4% (287) of them never felt discomfort in their hip/buttocks and very fewer SMOs felt comfort with their left foot (3.6%, n = 12) and left knee (7.4%, n = 25). The maximum number of the SMOs sense the discomfort in their foot (Left (29.2%, n = 98) and Right (27.1%, n = 91)) several times in a day and the severity (Left (35.4%, n = 119) and Right (32.4%, n = 109)) of the discomfort substantially interfered (Left (37.2%, n = 125) & Right (35.4%, n = 119)) with their work.

Table 8.4 shows the sectional score and total discomfort score with the percentage of different body parts of SMOs (n = 336). The frequency, severity, and interference score of the different body parts were calculated by using the Eq. (8.1) and the total discomfort score by Eq. (8.2). Finally, the discomfort score has presented in terms of percentage for the different body parts. Figure 8.5 shows the graphical representation of discomfort score (%) of different body parts. Based on the discomfort score percentage, the body parts arranged in descending order (Table 8.5).

8.3.3 REBA

The body posture of the participants during sewing tasks was observed in accordance with the REBA form and the scores were recorded. A separate REBA form was used for each operator and observation. The observations were carried out without disturbing the regular task of the operators. The operators' mean score A, score B, and grand score were found to be 5.4 (SD: 1.3, range: 3–8), 3.8 (SD: 1.3, range: 2–6) and 7.98 (SD: 2.2, range: 5–12) respectively (Table 8.6).

Score A included the trunk, neck and legs and score B included upper arms, lower arms, and wrists. The grand score includes the activity score too. The mean score A of 5.4 indicates that the operators' trunk, neck, and leg were in medium risk level and need to change soon. The risk scores of the legs and neck were determined to be within normal limits. But, due to standing operation the leg has to bear the stress created by the upper body. The score B found to be 3.8 ($\approx$4) indicates that the operators' upper arms, lower arms, and wrists were in medium risk level and need to change soon.

The final mean REBA score of 7.98 ($\approx$8) indicates that the operators' postures at high risk and their workstations need to be investigated and required changes are required immediately. There were no operators who received REBA scores of 1–3, which indicates an acceptable posture (Table 8.7). A maximum number of the

Table 8.3 Discomfort, severity and interference of SMOs with respect to different body parts (n = 336)

Body parts		Never (%)	1–2 times (%)	3–4 times (%)	Once every day (%)	Several times every day (%)	Slightly uncomfortable (%)	Moderately uncomfortable (%)	Very uncomfortable (%)	Not at all (%)	Slightly interfered (%)	Substantially interfered (%)
Neck		27.1	18.8	31.3	14.0	8.9	40.5	28.9	3.6	24.4	25.9	22.6
Shoulder	Left	19.0	8.6	24.4	26.5	21.4	36.0	31.0	14.0	22.6	32.7	25.6
	Right	17.9	6.8	26.5	28.3	20.5	35.1	30.4	16.7	21.1	32.1	28.9
Upper back		21.7	10.1	22.6	25.6	19.9	37.8	31.3	9.2	23.2	30.1	25.0
Upper Arm	Left	24.1	14.6	28.3	21.4	11.6	39.0	32.4	4.5	25.6	27.1	23.2
	Right	22.6	12.2	24.4	23.2	17.6	38.4	32.7	6.3	23.8	29.2	24.4
Lower back		16.4	6.3	25.9	29.5	22.0	35.4	35.7	12.5	19.0	34.8	29.8
Forearm	Left	33.3	24.4	32.4	6.0	3.9	41.7	20.8	4.2	14.3	27.1	25.3
	Right	28.9	21.4	30.4	10.7	8.6	40.5	22.3	8.3	16.4	31.3	23.5
Wrist	Left	37.8	27.1	26.2	5.7	3.3	42.3	17.0	3.0	11.9	26.5	23.8
	Right	35.4	24.4	27.4	7.4	5.4	41.4	18.5	4.8	12.8	29.8	22.0
Hip/buttocks		85.4	9.5	3.0	1.5	0.6	10.1	3.9	0.6	12.2	2.4	0.0
Thigh	Left	11.3	6.0	28.9	30.4	23.5	32.4	33.6	22.6	35.7	21.7	31.3
	Right	14.0	8.6	26.8	27.7	22.9	33.3	29.2	23.5	36.3	17.3	32.4
Knee	Left	7.4	10.4	24.7	32.4	25.0	30.1	29.2	33.3	31.3	28.0	33.3
	Right	9.5	14.6	25.0	29.2	21.7	32.7	27.1	30.7	32.4	22.3	35.7
Lower Leg	Left	9.2	11.3	24.1	31.5	23.8	31.3	35.4	24.1	32.4	24.7	33.6
	Right	11.6	17.6	22.9	27.4	20.5	33.6	30.4	24.4	31.5	24.4	32.4
Foot	Left	3.6	10.1	18.8	38.4	29.2	31.0	30.1	35.4	30.4	28.9	37.2
	Right	8.9	9.8	18.2	36.0	27.1	30.1	28.6	32.4	33.3	22.3	35.4

Table 8.4 Total discomfort score of different body parts

Body Parts		Frequency (2)	Severity (2)	Interference (2)	Discomfort score (1)	Discomfort score %
Neck		534	366	484	94594896	2.42
Shoulder	Left	748	470	554	194764240	4.99
	Right	762	490	578	215813640	5.53
Upper back		712	430	532	162877120	4.17
Upper Arm	Left	611	394	502	120848468	3.10
	Right	675	412	522	145168200	3.72
Lower back		788	485	598	228543640	5.85
Forearm	Left	412	322	485	64342040	1.65
	Right	500	370	502	92870000	2.38
Wrist	Left	368	286	458	48203584	1.23
	Right	413	311	465	59725995	1.53
Hip/buttocks		75	66	57	282150	0.01
Thigh	Left	836	563	581	273458108	7.00
	Right	796	545	565	245108300	6.28
Knee	Left	864	633	629	344007648	8.81
	Right	803	601	619	298731257	7.65
Lower Leg	Left	838	586	614	301515752	7.72
	Right	765	563	597	257124915	6.59
Foot	Left	939	663	671	417735747	10.70
	Right	882	620	619	338493960	8.67

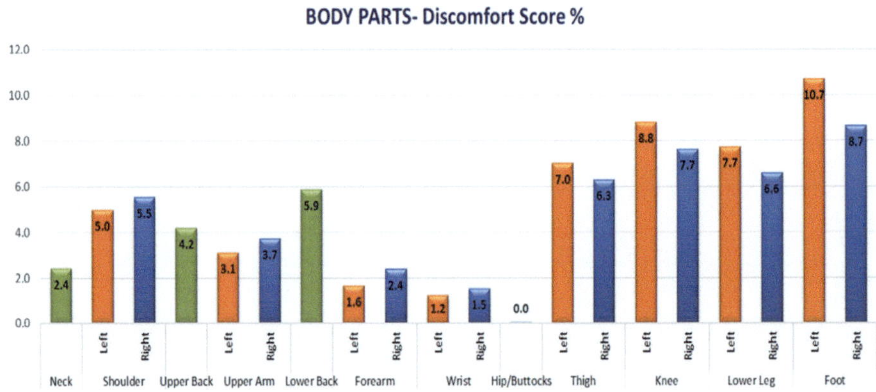

Fig. 8.5 Graphical representation of discomfort score (%) of different body parts

operators (Score A ~ 88%, Score B ~ 53% and Grand score ~ 49%) were in the medium risk level and action necessary. The graphical representation of the percentage of operators in different risk level of the REBA analysis shown in Fig. 8.6.

Table 8.5 Priority of body parts based on discomfort

Body parts		Discomfort score %	Ranking of body parts
Foot	Left	10.7	1
Knee	Left	8.8	2
Foot	Right	8.7	3
Lower leg	Left	7.7	4
Knee	Right	7.7	5
Thigh	Left	7.0	6
Lower leg	Right	6.6	7
Thigh	Right	6.3	8
Lower back		5.9	9
Shoulder	Right	5.5	10
Shoulder	Left	5.0	11
Upper back		4.2	12
Upper arm	Right	3.7	13
Upper arm	Left	3.1	14
Neck		2.4	15
Forearm	Right	2.4	16
Forearm	Left	1.6	17
Wrist	Right	1.5	18
Wrist	Left	1.2	19
Hip/buttocks		0.0	20

Table 8.6 Mean, SD and range of REBA score

	Neck	Trunk	Leg	Score A	Upper Arm	Lower Arm	Wrist	Score B	Grand score
Mean	2.3	3.2	1.9	5.4	1.9	2.0	1.8	3.8	7.98
SD	0.46	1.09	0.31	1.34	0.81	0.0	0.76	1.31	2.17
Range	2–3	2–5	1–2	3–8	1–3	2	1–3	2–6	5–12

Table 8.7 Risk level and action of the observed REBA score

REBA score	Risk level	Action	Score A		Score B		Grand score	
			n	%	n	%	n	%
1	Negligible	None Necessary	0	0	0	0	0	0
2–3	Low	May Be Necessary	22	6.55	158	47.02	0	0
4–7	**Medium**	**Necessary**	**297**	**88.39**	**178**	**52.98**	**165**	**49.11**
8–10	High	Necessary Soon	17	5.06	0	0	109	32.44
11–*15*	Very High	Necessary NOW	0	0	0	0	62	18.45

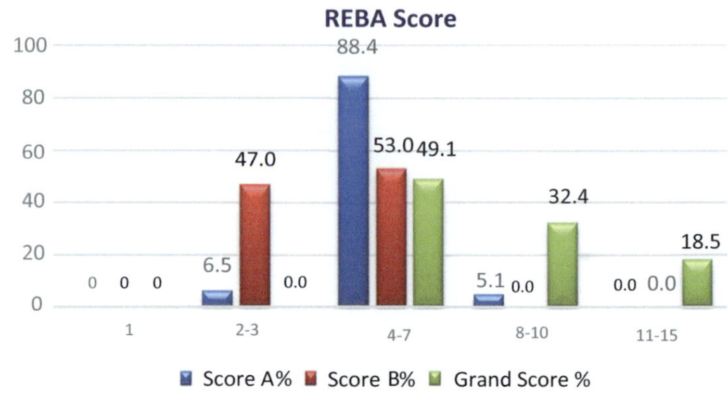

Fig. 8.6 Graphical representation of percentage of operators in REBA risk level

8.4 Discussion

This study conducted to determine MDSs and ergonomic risks in female SMOs is the first of standing modular sewing production in Sri Lanka. Women are the principal workers in the textile sector. The result of the demographic and job characteristics survey presented that the factors like age, education, marital status, and experience in MSP were not significantly affected with the operators discomfort. But, the factors like BMI, daily walking distance to boarding place or company, job satisfaction, and job stress were significantly influenced with operators' discomfort. In several studies, psychosocial factors, such as high job strain, less job satisfaction and to feel less passionate about work, have been reported to increase the MSD prevalence (David 2005; Denis et al. 2008; Choobineh et al. 2009). The similar studies of sitting sewing operators' discomfort were significantly influenced by the factors like medical history of systemic illness, use of scissors and work pressure, but BMI was not a significant factor (Öztürk and Esin 2011). The results of the present study provide further evidence that both physical and psychosocial job factors are associated with increased discomfort among workers involved in sewing activities.

The CMDQ discomfort survey result presented that the influence of discomfort among the different body parts of the operators (Table 8.5). A maximum number of operators have felt more discomfort in their foot, knee, lower leg, thigh, lower back, and shoulder than upper back, upper arm, neck, forearm, wrist and hip/buttocks. Comparatively, the operators' discomfort was more in left than right in the body parts of foot, lower leg, knee, and thigh but reverse in shoulder, upper arm, fore arm, and wrist. The left leg bear the total weight of the body and most of the repetitive activities carried out by right hand with flexed left side. In the present study in the lower body more discomfort was felt than upper body. But in the sitting sewing operation, upper body gets more strain than lower body. The sitting operation studies reported that the overall prevalence of complaints was surprisingly low

(15.5%), with the back (8.9%) and knee (4.9%) the most commonly noted sites of pain (Lombardo et al. 2012) and high in neck, shoulder and back (Nazari et al. 2012; Roquelaure 2012). Öztürk and Esin (2011), reported that the areas identified by most women as complaint more discomfort than lower body (trunk (62.5%, 34.8% upper back, 23.9% lower back), neck (50.5%), shoulders (50.2%), and upper limbs (22.3%)). These findings provide evidence that more discomfort in lower body than upper body in standing sewing operation.

In this study, the operator's ergonomic risks were determined by REBA scores and presented the result (Table 8.6) with SD and range of the score with respect to different body parts, score A and B, and grand score. The leg (REBA) score was below low risk level but felt maximum discomfort (CMDQ). The leg score indicates that the discomfort was not due to awkward posture and due to stress created by upper body weight. The score A (Range-3–8, Mean-5.4) indicates that the 88.39% of the operators' trunk, leg, and neck were in medium risk level and necessary to change. The score B (2–6, 3.8) indicates that the 52.98% of the operators' upper arm, lower arm and wrist were also in medium risk level and necessary to change. Finally, the grand score (5–12, 7.98) indicates that the 49.11% of the operators were in medium risk level, 32.44% were in high risk level and need to change soon and 18.45% of operators were in high risk level (Tables 8.7). The most of the operators were in medium risk level in the MSP operators in lean environment.

8.4.1 Managerial Implications

The study exposed to management that the operators' ergonomics risks level and interference with work in the lean environment sewing production floor. The posture analysis exposed that the most of the operators were in medium risk level in the floor. The study motivates that the awareness of the ergonomic aspects in MSP floor. The result of the study motivates the management to provide the necessary action and resources to overcome the ergonomic issues in the floor. Once the management commitment ensures the development project cost, the development projects can be initiated in the MSP floor. An initial investment has to be made by management to provide resources. Once the development projects were done, a study can compare the ergonomics benefits in terms of cost.

8.5 Conclusions

The CMDQ survey indicates that the high occurrence of musculoskeletal ache and discomfort (mostly in the foot, knee, lower leg, thigh, low back, and shoulders) among the standing sewing machine operators emphasizes the need for ergonomic interventions for improving the working conditions of the lean environment. The relatively medium REBA scores in the study highlight that the operators were in medium risk level and the work station redesign and ergonomic awareness programs. The demographic characteristics like BMI and daily walking distance were a significant factor for MSD and job characteristics factors including job stress and job satisfaction were also found to be associated with the occurrence of musculoskeletal symptoms in different body regions. The ergonomic solutions like fixing the proper height of sewing machine, inclined work table, providing anti-fatigue mats, job rotation, stretching exercises and short-break approach is suggested for minimizing the absenteeism and turnover due to body ache and to increase the efficiency and quality of the product and process.

There is further research scope to consider more demographic and job characteristics and more levels in discomfort answering part. Comparative study can between standing and sitting operation of sewing production among female operators, and before/after ergonomic development in terms of performance metrics.

References

Aase GR, Olson JR, Schniederjans MJ (2004) U-shaped assembly line layouts and their impact on labor productivity: an experimental study. Eur J Oper Res 156:698–711

Andersson ER (1992) Economic evaluation of ergonomic solutions: Part I—Guidelines for the practitioner. Int J Ind Ergon 10:161–171

Bhamu J, Sangwan KS, Singh Sangwan K (2014) Lean manufacturing: literature review and research issues. Int J Oper Prod Manag 34:876–940

Brisson C, Vinet A, Vezina M (1989) Disability among female garment workers: a comparison with a national sample. Scand J Work Environ Heal 15:323–328

Buckle PW, Jason Devereux J (2002) The nature of work-related neck and upper limb musculoskeletal disorders. Appl Ergon 33:207–217

Choobineh A, Sani GP, Rohani MS, Pour MG, Neghab M (2009) Perceived demands and musculoskeletal symptoms among employees of an Iranian petrochemical industry. Int J Ind Ergon 39:766–770

Dahlberg R, Karlqvist L, Bildt C, Nykvist K (2004) Do work technique and musculoskeletal symptoms differ between men and women performing the same type of work tasks? Appl Ergon 35:521–529

David GC (2005) Ergonomic methods for assessing exposure to risk factors for work-related musculoskeletal disorders. Occup Med (Lond) 55:190–199

Denis D, St-Vincent M, Imbeau D, Jetté C, Nastasia I (2008) Intervention practices in musculoskeletal disorder prevention: a critical literature review. Appl Ergon 39:1–14

Dianat I, Kord M, Yahyazade P, Karimi MA, Stedmon AW (2015) Association of individual and work-related risk factors with musculoskeletal symptoms among Iranian sewing machine operators. Appl Ergon 51:180–188

Gallagher S, Schall MC Jr (2016) Musculoskeletal disorders as a fatigue failure process: evidence, implications and research needs. http://dx.doi.org/10.1080/00140139.2016.1208848. 139, 0

Hignett S, McAtamney L (2000) Rapid Entire Body Assessment (REBA). Appl Ergon 31:201–205

Jansen K, Luik M, Reinvee M, Viljasoo V, Ereline J, Gapeyeva H, Pääsuke M (2012) Musculoskeletal discomfort in production assembly workers. Acta Kinesiol Univ Tartu 18:102–110

Kaergaard A, Andersen JH (2000) Musculoskeletal disorders of the neck and shoulders in female sewing machine operators: prevalence, incidence, and prognosis. Occup Environ Med 57:528–534

Kilroy N, Dockrell S (2000) Ergonomic intervention: Its effect on working posture and musculoskeletal symptoms in female biomedical scientists

Kogi K, Kawakami T, Itani T, Batino JM (2003) Low-cost work improvements that can reduce the risk of musculoskeletal disorders. Int J Ind Ergon 31:179–184

Krijenen RM, De Bore E, Ader H, Bruynzeel D (1997) Venous insufficiency in male workers with a standing profession. Part: 1 Epidemiology. Dermatology 194:111–120

Li G, Haslegrave CM, Corlett EN (1995) Factors affecting posture for machine sewing tasks. The need for changes in sewing machine design. Appl Ergon 26:35–46

Lombardo SR, Vijitha de Silva P, Lipscomb HJ, Østbye T (2012) Musculoskeletal symptoms among female garment factory workers in Sri Lanka. Int J Occup Environ Health 18:210–219

Nazari J, Mahmoudi N, Dianat I, Graveling R (2012) Working conditions in carpet weaving workshops and Muscu-loskeletal complaints among workers in Tabriz-Iran. Heal Promot Perspect 2:265–273

Oh I-H, Yoon S-J, Seo H-Y, Kim E-J, Kim YA (2011) The economic burden of musculoskeletal disease in Korea: a cross sectional study. BMC Musculoskelet. Disord. 12:157

Oksa J, Ducharme MB, Rintamäki H (2002) Combined effect of repetitive work and cold on muscle function and fatigue. J Appl Physiol 92:354–361

Öztürk N, Esin MN (2011) Investigation of musculoskeletal symptoms and ergonomic risk factors among female sewing machine operators in Turkey. Int J Ind Ergon 41:585–591

Piedrahíta H, Punnett L, Shahnavaz H (2004) Musculoskeletal symptoms in cold exposed and non-cold exposed workers. Int J Ind Ergon 34:271–278

Piranveyseh P, Motamedzade M, Osatuke K, Mohammadfam I, Moghimbeigi A, Soltanzadeh A, Mohammadi H (2016) Association between psychosocial, organizational and personal factors and prevalence of musculoskeletal disorders in office workers. Int J Occup Saf Ergon 22:267–273

REBA-Worksheet-v-2.0.pdf, http://ergo-plus.com/wpcontent/uploads/REBA.pdf

Roquelaure Y, Mariel J, Fanello S, Boissière J-C, Chiron H, Dano C, Bureau D, Penneau-Fontbonne D (2002) Active epidemiological surveillance of musculoskeletal disorders in a shoe factory. Occup Environ Med 59:452–458

Sealetsa OJ, Thatcher A (2011) Ergonomics issues among sewing machine operators in the textile manufacturing industry in Botswana. Work 38:279–289

Shah R, Ward PT (2007) Defining and developing measures of lean production. J Oper Manag 25:785–805

Spasojević Brkić VK, Klarin MM, Brkić AD (2015) Ergonomic design of crane cabin interior: the path to improved safety. Saf Sci 73:43–51

Widanarko B, Legg S, Devereux J, Stevenson M (2014) The combined effect of physical, psychosocial/organisational and/or environmental risk factors on the presence of work-related musculoskeletal symptoms and its consequences. Appl Ergon 45:1610–1621

Zander JE, King PM, Ezenwa BN (2004) Influence of flooring conditions on lower leg volume following prolonged standing. Int J Ind Ergon 34:279–288

Chapter 9
Ergonomic Design and Development of Material Handling Equipment for Pantry Car Personnel

Abhijeet Arvind Raut and Jyotiraman De

Abstract Proper working method and equipment may play a very important role in increased productivity and reduced fatigue. This paper presents Ergonomic design and development of a novel material handling equipment (trolley) for Indian Railways pantry car personnel. Pantry car personnel are seen carrying a heavy stack of plates for long distances with many twists and turns because of which they may have enormous physical stress and fatigue causing Musculoskeletal Disorders (MSDs). In order to address the problem of MSDs among the pantry car personnel, a trolley is designed requiring less floor space, following ergonomic principles and using anthropometry data. A number of design parameters, such as weight and collapsible structure dimension of railway coach and weight of each plate, are considered in the improved design of the trolley. A prototype has been developed and tested for 10 kg load application. This paper presents field study, surveys, analysis of data collected, innovative product development, features, and results and discussion of the trolley.

Keywords Human factors · Ergonomics · Anthropometry · Prototypes · Product development

9.1 Introduction

In many industries and workplaces material handling is a common activity. Lifting, an activity of manual material handling, is considered as the most stressful activity and often affects employees and results to costly occupational injuries to the

A.A. Raut (✉) · J. De
Department of Mechanical Engineering, G. H. Raisoni College of Engineering (Autonomous), Nagpur, Maharashtra, India
e-mail: abhijeetraut21@gmail.com

P.K. Ray and J. Maiti (eds.), *Ergonomic Design of Products and Worksystems—21st Century Perspectives of Asia*, Managing the Asian Century,
https://doi.org/10.1007/978-981-10-5457-0_9

company and is one of the major concerns in today's industries (Dagenais et al. 2008; Faber et al. 2009). Other activities such as pushing, pulling, and carrying, if done improperly may lead to injuries. These injuries are often termed as musculoskeletal disorders. It affects the quality of work, work time, and productivity which in turn affects economy of employee, organization, and society (Mital and Ramakrishnan 1999; Andersson 1992). These disorders are more common among young or healthy workers indicating the faulty work practices and equipment designs. One such area is catering industry or food material handling where the need of material handling equipment was understood and developments were made.

Since the invention of shopping carts by Sylvan Goldman on June 4, 1937, the trend of lightweight material handling has changed a lot. Today varieties of trolleys are available according to the application. In Indian Railway trains, pantry car is provided to serve the passengers with food. Pantry car personnel are deployed for serving the food throughout the train. But generally, the personnel experience fatigue during work while transferring and serving food to the passengers. Many reasons for fatigue can be heavyweight being carried for long distances, personnel need to twist and turn during passing from one coach to another or presence of crowd and/or improper load handling. As per the rules of railway that the railway compartment should contain only the reserved passengers, some specific trolleys can be used effectively. But considering the regular condition of compartment where it is overcrowded by the passengers, luggage, and vendors, no such trolley exist which will travel smoothly through the crowd and also reduce the physical stress of personnel. In order to resolve the above issues, attempt is required to be made for design and development of material handling equipment or trolley considering human factors.

Many types of trolleys are available in the market for various uses. Most widely used trolley is shopping trolley. For industrial purposes, similar trolleys are used with required modification. A few dedicated trolleys are also in use for catering purpose but are not suitable for regular express and superfast trains of Indian Railways. Considering the Indian railway coach conditions, the disadvantages of these are as follows:

- Floor space required is more
- High movability as required in railway coaches not possible
- Not designed considering ergonomic and anthropometric for use in coaches
- Height adjustability not provided
- Not foldable.

In India as a developing country, working methods and conditions are not proper and also injuries like musculoskeletal disorders due to manual material handling are very serious. This paper attempts to address one such issue and contributes to fill the gap.

9.2 Literature Survey

For designing any material handling equipment, many factors are considered such as human factors (height, weight, age, gender, strength, posture, physiological capacity), task factors (distance moved, movement initiation force requirements, sustained motion force requirements, direction and nature of movement, duration of pushing/pulling task), cart/equipment factors (handhold height, handhold orientation, handhold type, caster/wheel design specifications, stability, size, weight), floor/ground factors (surface characteristics, slope, contaminants) (The Ergonomics of Manual Material Handling-Pushing and pulling tasks 2001). In this time of rapid development companies cannot afford a loss in terms of less productivity and manpower losses due to MSD's caused by bad work practices and faulty equipment (Colombini et al. 2001).

9.2.1 Musculoskeletal Disorders (MSDs)

Improper work posture or wrong work practices lead to various injuries or disorders that affect human body and cause temporary or sometimes permanent injury to musculoskeletal system (Ergonomics Guidelines for Manual Handling 2010). Medical term for these common and costly problems for people is given as Musculoskeletal Disorders (MSDs). The ergonomic factors causing musculoskeletal disorders in pantry car personnel are given in the following:

High task repetition: Repetitive tasks needed to be carried out by the personnel to meet their service target.
Forceful exertions: Carrying stacks of trays imparts high force loads on personnel body. Due to this muscle force increases along with the increase in associated fatigue which may lead to MSD (Introduction to Ergonomics 2015).
Repetitive or sustained awkward postures: Improper postures put excessive force on joints and effects the muscles and tendons around them. If the same posture is maintained for a longer duration, it increases the risk of MSD (Introduction to Ergonomics 2015).
Poor workpractices: Improper work culture and lifting techniques also increases the risk of MSDs, fatigue and reduces body's recovering capability.

Wrong or improper body postures and work practices leads to physical stresses. The discomfort caused to the body specifically in muscle or joints are indicators to wrong workpractices. Repetitive actions of gripping and position often lead to symptoms such as chronic pain, swelling, or painful joints. Musculoskeletal Disorders affect the various body parts such as neck, shoulders, wrists, back (upper and lower), hips, legs, knees, and feet. As the human body consists of various musculoskeletal systems, various other disorders can produce significant musculoskeletal symptoms such as low back pain, fibromyalgia, gout, different types of arthritis and tendinitis. Some of these disorders can cause mild discomfort to debilitating pain (Musculoskeletal Disorders 2015). Low back pain is the most common MSD.

9.2.2 Ergonomics

A very popular term among industries, product design companies and researchers is ergonomics. Ergonomics is all about man–machine relationship. It is a science focused on the study of human fit, and decreased fatigue and discomfort through product design. Many common activities carried out in industries, stores and service places are lifting, lowering, pulling, pushing, carrying loads and moving (Ergonomics in the Workplace 2006). Lifting and lowering is one of the major activities being carried out among all. Figure 9.1 shows the basic numerical limits for lifting and lowering. This guideline of weight limits can be applied if load can be easily handled and working environment and body is stable. If the hands enter more than one of the box zones during the task, then the smallest weight limit should be used.

- **Frequency**: The numerical weight limits are considered for the tasks which are done less than 30 times an hour. The weight limits should be reduced by 30, 50, and 80% if the operation is repeated once or twice a minute, 5–8 times per minute, and 12 times a minute respectively.
- **Twisting**: While twisting by 45° and 90°, the handled weights should be lowered by about 10 and 20%, respectively.
- **Carrying**: The load should not be carried more than 10 m without stopping to rest (Introduction to Ergonomics 2015).

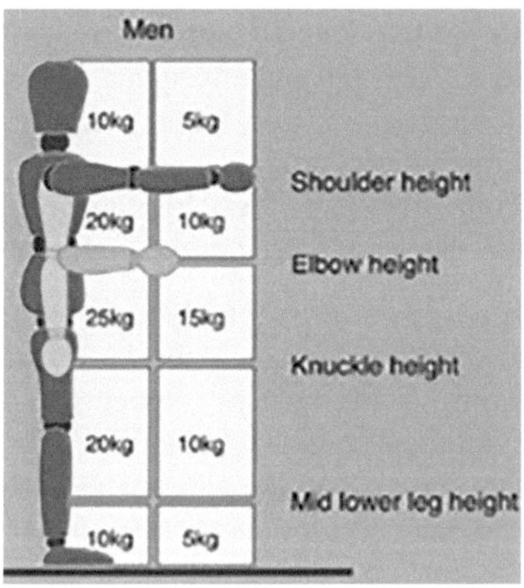

Fig. 9.1 Numerical limits for lifting and lowering (Ergonomics in the Workplace 2006) Also, the above numerical limits vary by the following factors (Ergonomics in the Workplace 2006)

Pushing and pulling is another activity which often causes internal strain and may lead to MSD's (Malchaire et al. 1996). One of the general and most common solutions to these problems are the use of material handling equipment such as trolley. Stress on back and shoulders reduce drastically when pushed or pulled when compared to lifting (Schibye et al. 2001). But trolley should be designed considering various factors. A trolley should be light in weight, easily handled and stored. Handle height should be proper to reduce the mechanical effort needed by the user and reduce strain (Kuiper et al. 1999; Kuijer et al. 2007). For minimum effort to push or pull, the handle height should be placed between the elbow and hip (Report by Government of the Canadian Province of Alberta 2004). It is suggested that while pushing force exerted by hands should be away from the body and parallel to the motion direction to reduce the efforts by 10% compared to holding perpendicularly (Hoozemans et al. 1998; Seo et al. 2010). Larger wheels require less force and are faster, stable in narrow slot and rough surfaces (Malchaire et al. 1996; Young et al. 1997). Use of castor wheel reduces turning effort and eliminates the need of space for turning (Measuring Push/Pull Forces on Carts 2004).

9.2.3 Anthropometry

With reference to the Indian Anthropometric data (Indian Anthropometric dimensions for ergonomic design practice 2015), the elbow and waist height of males, females and combined data were obtained as shown in Table 9.1.

The workstation or handling equipment design will have a major effect on the working postures. Optimal working height varies with the load to be handled as shown in Table 9.2. As the highest weight carrying capacity is between the elbow and waist height, heavy work should be done in this range whereas, the light and precision work can be done at approximately elbow height. Ideally, a workstation or handling equipment should be designed to fit a wide range of employees (smallest, tallest, and average). Increasing the handling equipment adjustability is the best way to promote good working condition (Ergonomics Guidelines for Manual Handling 2010).

There were many literature works found on ergonomics, MSD, and industrial material handling equipment but the literature on trolleys meant for low floor space conditions and crowded places such as in crowded railway coach or emergency halls are not available.

Table 9.1 Measurement as per Indian Anthropometric Data

	Males	Females	Combined
Elbow Height	1038	956	1018
Waist Height	972	931	963

Table 9.2 Optimal working height is based on worker's body dimension

Sr. No.	Work Type	Optimal work height
1	Heavy Work	Between elbow and waist height
2	Light Work	At approximately elbow height
3	Precision Work	At approximately 5 cm above elbow height

9.3 Field Survey and Analysis

Crowded railway coach is one suitable example for low floor space and crowded place. We have considered meal delivery by railway pantry personnel for development of trolley in such conditions. In order to understand the concerns of the personnel, a survey was conducted on 56 pantry car personnel of Indian Railway pantry cars. The majority of population was male with age ranging from 21 to 35 years. The data has been gathered through questionnaires and visual inspection. The questionnaires were used to identify the problems faced by the personnel and recommendations from them to solve the problems. Other than this surveys for length of Indian Railways coaches and weight of each food plate were also conducted to understand the working condition and environment properly.

9.3.1 Indian Railway Coaches

There are various types of coaches used in Indian railway having different design and dimensions. As per the data from the Indian Railway sources there are two major types of coaches used in Indian Railway:

ICF: Integral Coach Factory, Perambur (Chennai) and Rail Coach Factory, Kapurthala (Punjab)
LHB Coaches: LHB body with FIAT bogie (Linke Holfmann Busch—German) Maximum Permissible Speed 160 kmph.

Considering the buffer to buffer length of ICF/RCF coaches as shown in Table 9.3, to cross one coach personnel needs to cover 22.296 m. If we consider 16 reserved coach train, pantry car is generally placed at middle. So, if personnel need to serve each coach one side, he needs to travel

$$22.296 * (8+7+6+5+4+3+2+1) = 802.656\,\text{m}.$$

Table 9.3 Length of Indian Railway coaches

	ICF/RCF (m)	LHB/HYBRID (m)
Length over buffer	22.296	24.00
Length over head stock	21.336	23.54

Table 9.4 Menu card with weight details

Sr. No.	Menu	Weight (kg)
1	Rice Pulao or Jira Rice or plain Rice of fine quality	0.150
2	Paratha (2 nos) or Chapati (4 nos) or Poories (5 nos)	0.100
3	Dal or Sambhar (Thick consistency)	0.150
4	Mixed vegetable (seasonal)	0.100
5	Curd—100 g or sweet	0.040
6	Pickle in sachet	0.015
7	Packaged drinking water in sealed glass—300 ml	0.300
Total Weight		0.855

9.3.2 Weight Carried by Pantry Car Personnel

Data obtained from the Indian Railway Catering and Tourism Corporation Ltd. website (Menu for Standard Breakfast and Standard Meals—Mail/Express Trains 2015) on weight of food items on each plate is shown in Table 9.4.

It is observed that the stacks of plates are carried by single personnel. On an average stack of plates carried by the personnel consists of 12 plates at a time. As per the data obtained from Table 9.4.

If 12 plates are to carry by single personnel:

Total Weight to carry = 10.26 kg
Average distance to be covered by pantry car personnel carrying 10.26 kg load = 802.656 m
Sometimes the load is shared by two personnel. One carries curd and packaged drinking water whereas other carries rest of the food on plates.

If curd and packaged drinking water handled by second person:

For first person, total weight to carry = 6.18 kg
For second person, total weight to carry = 4.08 kg
But in the above case manpower requirement increases, which is not desirable as the manpower requirement increases and cost to company increases.

9.3.3 Present Working Condition

Length of a single coach is 22.296 m and it is covered at a stretch without stopped by the pantry car personnel which is more than the suggested 10 m limitation. Also the position of hand holding stack of trays remains same throughout the travel in spite of many twists and turns. Hence, the load handled in a regular manner by the personnel is ergonomically risky and can cause fatigue leading to loss in efficiency and production. The actual condition of railway pantry car personnel can be understood by the following Fig. 9.2.

9.3.4 Problems Faced by the Catering Personnel

For the pantry car personnel, the root cause of MSDs is exposure to MSD risk factors—both work-related risk factors and individual-related risk factors. These are mainly caused due to carrying heavy loads for long distance and continuous body movement while carrying the load. Both of these risk factors are experienced by the

(a) **(b)**

Fig. 9.2 Actual food plate handling conditions in **a** Howrah Ahmedabad Express (Date of picture taken 10-06-15) (*Left side*) **b** Bangalore Special Superfast (Date of picture taken 23-09-15) (*Right side*)

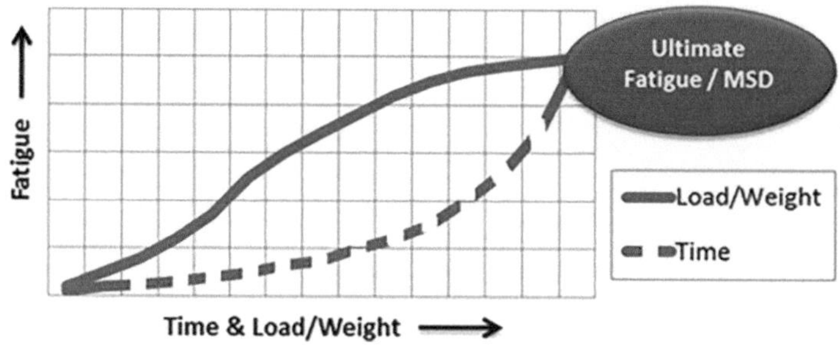

Fig. 9.3 Fatigue versus Time and Load/Weight Curve

pantry car personnel while serving the lunch plates from pantry car to the particular seat. As the duration of such work increases the fatigue level also increases as shown in Fig. 9.3.

9.4 Product Development Considering Ergonomics and Anthropometric Data

Keeping the ergonomics in mind, a product outlook was generated for basic idea and other dimensions were decided for the trolley.

9.4.1 Product Outlook and Description

If the regular condition of railway compartment is considered, it is filled with passenger and their luggage. Also, the area for passage is less compared to the area needed for the trolleys used in mall or industrial shop floor. In order to provide a trolley system in such application, it is required to design a trolley having less floor space requirement. Outlook of such a trolley is generated which can be illustrated using Fig. 9.4.

9.4.1.1 Holding Platform and Handle Height

Considering the ergonomics and anthropometry data, lowest and lowest height of waist and highest height of elbow the range of height adjuster is decided. Hence the decided minimum and maximum height provided to the product is 931 and 1038 mm respectively.

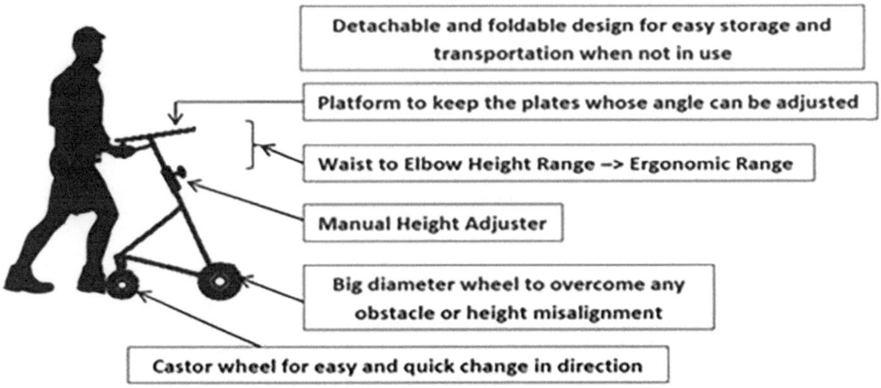

Fig. 9.4 Brief product outlooks

9.4.1.2 Height Adjusting Mechanism

The height of the product is to be kept up to the waist. The common waist height is considered from the Indian Anthropometric data. A long rod is provided in order to join the wheel base and the top plate. It also provides the required height to the top plate. A telescopic design is used with knob setting to fix the columns after adjusting the height. This is achieved by matching the knob (fixed on lower column) with the holes provided in the upper column.

9.4.1.3 Hands-Free Ability System

When the personnel will serve the plates to the customer, he needs to release his hands from the top plate. In this condition the balance will be lost. To maintain the balance and stability the carrier is provided with a foldable foot balancer which can be engaged just with the simple effort of leg.

9.4.2 Designs to Product—Concept to Creation

The idea and outlook of the product was converted into ergonomic material handling equipment for Indian Railways pantry car personnel. After designing the trolley, prototype was developed as shown in Fig. 9.5.

The designed trolley is lightweight, durable, and collapsible in structure. This results in easy handling by pantry car personnel and proper storage of the trolley when not in use.

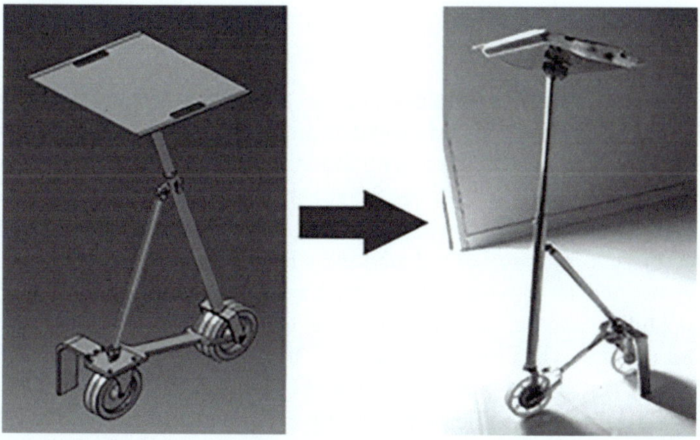

Fig. 9.5 Design to Prototype

9.5 Results and Discussion

After development of prototype, test was carried out to confirm the working of the prototype. The prototype was able to carry load up to 20 kg without any problem. All the targets were achieved by the developed material handling equipment. Stainless steel is used as material for development. After design calculation and also according to particular condition at panty like Heat, load, crowd, etc., due to which rusting can take place on parts of product so that S.S. is used, also it is lightweight and robust for ease of handling. Force to be exerted by the personnel was reduced from 200 to 10 N by replacing lifting activity by pushing activity. Trolley reduced the load by a significant amount to be handled by the personnel leaving only the balancing part to him by transferring the load to the base. Trolley was balanced after engaging the foot balancer providing hands-free condition to the personnel. This would enable the personnel to serve the food and take money from the customer without any difficulty. Also, quick maneuverability during turns and from in between the crowd was possible due to implementation of castor wheel. Overall the trolley was able to reduce the stress of the personnel. Also, the trolley was collapsible which took less space for storage.

The use of trolley would prevent early fatigue to the personnel, improve morale, work efficiency, reduce manpower requirement, and cost to company because of occupational injuries. For more improved and robust version the following advancement can be done:

- Built-in battery and motor assisted rear wheel.
- Replaceable top plate as per application; e.g., catering, stores.
- Stainless steel body can be replaced with plastic body for light weight.
- Present foot balancer system can be replaced with a more robust one.

9.6 Conclusions

Proper equipment and workpractices are the key to less fatigue and higher productivity. For low floor space and crowded areas like railway catering, which also consists of high load transfer for long duration with frequent twists and turns, violate the ergonomics and lead to fatigue to the personnel. The product was designed and developed considering the floor space requirement, ergonomics, and anthropometry data. The trolley reduced the stress to the personnel by reducing the force to be exerted to move the load, enabled quick maneuverability, high load carrying capability, and hands-free conditions were also achieved. This satisfied the purpose and promotes healthy work practice by reducing the musculoskeletal disorders. The product can be modified to make it more advanced and suitable for multipurpose use. With modifications on upper platform, it can be effectively used for other applications such as hospitals, transportation, and crowded areas.

References

Andersson E (1992) Economic evaluation of ergonomic solutions: Part I—Guidelines for the practitioner. Int J Ind Ergon 10(1–2):161–171

Colombini D, Occhipinti E, Delleman N, Fallentin N, Kilbom A, Grieco A (2001) Exposure assessment of upper limb repetitive movements: a consensus document developed by the Technical Committee on Musculoskeletal Disorders of International Ergonomics Association (IEA) endorsed by International Commission on Occupational Health (ICOH). G Ital Med Lav Ergon, 129–142

Dagenais S, Caro J, Haldeman S (2008) A systematic review of low back pain cost of illness studies in the United States and internationally. Spine J 8(1):8–20

Ergonomics Guidelines for Manual Handling (2010) WorkSafeNB, N.B. Canada, 4th edn. p 10

Ergonomics in the Workplace (2006) Health and safety authority, Dublin. ISBN 1-84496-036-6

Faber G, Kingma I, Bakker A, van Dieën J (2009) Low-back loading in lifting two loads beside the body compared to lifting one load in front of the body. J Biomech 42(1):35–41

Government of the Canadian Province of Alberta (2004) Workplace health and safety. Occup Health Saf Mag: Ergon, 1–4

Hoozemans M, Van Der Beek A, Fringsdresen M, Van Dijk F, Van Der Woude L (1998) Pushing and pulling in relation to musculoskeletal disorders: a review of risk factors. Ergonomics 41 (6):757–781

Indian Anthropometric dimensions for ergonomic design practice (2015) [Online] Available at: http://iitg.vlab.co.in/?sub=72&brch=171&sim=829&cnt=1530. Accessed 02 Dec 2015

Indian Railway, India (2015) Menu for standard breakfast and standard meals—Mail/Express Trains. Indian Railway Catering and Tourism Corporation Ltd., Delhi, p 298

Introduction to Ergonomics (2015) [Online] Available at: https://vula.uct.ac.za/access/content/group/9c29ba04-b1ee-49b9-8c85-9a468b556ce2/DOH/Module%201%20(OH)/Ergonomics/Ergo-9.html. Accessed 08 Sept 2015

Kuijer P, Hoozemans M, Frings-Dresen M (2007) A different approach for the ergonomic evaluation of pushing and pulling in practice. Int J Ind Ergon 37(11–12):855–862

Kuiper J, Burdorf A, Verbeek J, Frings-Dresen M, van der Beek A, Viikari-Juntura E (1999) Epidemiologic evidence on manual materials handling as a risk factor for back disorders: a systematic review. Int J Ind Ergon 24(4):389–404

Malchaire J, Cock N, Robert A (1996) Prevalence of musculoskeletal disorders at the wrist as a function of angles, forces, repetitiveness and movement velocities. Scand J Work Environ Health 22(3):176–181

Mital A, Ramakrishnan A (1999) A comparison of literature-based design recommendations and experimental capability data for a complex manual materials handling activity. Int J Ind Ergon 24(1):73–80

Musculoskeletal Disorders (2015) [Online] Available at: http://www.healthline.com/health/musculoskeletal-disorders. Accessed 25 Sept 2015

Schibye B, Sogaard K, Martinsen D, Klausen K (2001) Mechanical load on the low back and shoulders during pushing and pulling of two-wheeled waste containers compared with lifting and carrying of bags and bins. Clin Biomech 16(7):549–559

Seo N, Armstrong T, Young J (2010) Effects of handle orientation, gloves, handle friction and elbow posture on maximum horizontal pull and push forces. Ergonomics 53(1):92–101

State Fund CA (2004) Measuring push/pull forces on carts. [Online] Available at: http://www.scif.com/safety/ergomatters/Carts.html. Accessed 7 Nov 2015

The Ergonomics of Manual Material Handling-Pushing and pulling tasks (2001) Darcor Casters in cooperation with Ergo web, Ontario

Young S, Brogmus G, Bezverkhny I (1997) The forces required to pull loads upstairs with different hand trucks. In: Human Factors and Ergonomics Society 41st Annual Meeting. Human Factors and Ergonomics Society, Santa Monica, pp 697–701

Chapter 10
Study on Work-Related Musculoskeletal Disorders among Coir Industry Workers in the State of Kerala, India

M. Satheeshkumar and K. Krishnakumar

Abstract The aim of this study is to investigate the prevalence of work-related musculoskeletal disorders (WMSDs) among coir industry workers in the state of Kerala, India and to assess the demographic and physical factors (heat, noise, light, etc.) associated with such disorders. A cross-sectional study by a detailed questionnaire on symptoms of socio-demographic factors, musculoskeletal disorders, working postures and work history was administered to 362 subjects. The subjects are randomly selected from 12 coir mat manufacturing industry. A modified Nordic Musculoskeletal Questionnaire was used for the survey. The data collected were analysed by SPSS V20.0 package. Past 12-month prevalence of WMSDs at nine body regions were studied. The result shows that the lower back is the most prevalent region of WMSD disorders (58.8%). Significant relationships were existing to the variables gender, age, duration of employment, section of work and BMI with WMSD symptoms in all or certain body region. Also, it was identified that significant relationships existed to the physical factors such as heat, noise, dust and odour except light with WMSDs.

Keywords Coir industry · Prevalence · Work-related musculoskeletal disorders

10.1 Introduction

A work-related musculoskeletal disorders (WMSDs) is one of the major problems in industrialised countries and a major cause of disability (Nunes and Bush 2012). The musculoskeletal system provides form, support, stability and movement to the

M. Satheeshkumar (✉) · K. Krishnakumar
Department of Mechanical Engineering, College of Engineering Trivandrum, Thiruvananthapuram, Kerala, India
e-mail: msatheeshkumarm@gmail.com

K. Krishnakumar
e-mail: kkk@cet.ac.in

P.K. Ray and J. Maiti (eds.), *Ergonomic Design of Products and Worksystems—21st Century Perspectives of Asia*, Managing the Asian Century,
https://doi.org/10.1007/978-981-10-5457-0_10

body. It is made up of bones, muscles, cartilage, tendons, nerves, ligaments and other connective tissues. Work-related musculoskeletal disorders are injuries that affect the above system. Symptoms of WMSDs may include pain, discomfort, numbness and tingling in the affected area (Putz-Anderson et al. 1997). WMSDs are the most common work-related health problem and biggest causes of absence among millions of workers across the member state in Europe. In some states, 40% of the costs of workers' compensation are caused by WMSDs and up to 1.6% of the GDP of the country itself, it reduces company profitability and also add to the government's social costs (Podniece et al. 2008). In a report regarding WMSDs by Australian safety and compensation council, industries with the highest number of cases of WMSDs were from manufacturing, construction, retail trade, transport and storage, and health and community services sectors (Australia Safety Council 2006). WMSDs are found due to poor working postures, lack of job rotation, ergonomically poor design of work stations, poor design of plant layout, low salaries, absence of training, etc. Proper ergonomic design achieved through a thorough evaluation of workstation is required for reduced WMSDs.

Coir industry is the traditional industry in India, especially in Kerala state. The coir industry sector in Kerala provides direct employment to approximately 0.375 million people (Kerala Economic Review 2015). It contributed around 240 million US$ worth of exports during 2014–2015 financial year. Coir products from India are exported to 105 countries (Kerala Economic Review 2015). Coir industry workers adopt awkward postures, reparative movements, quick and fast movements of limbs, forceful exertion which are the important factors that affect their poor working efficiency and musculoskeletal disorders. Most of the coir industries are using traditional equipment and tools. Mechanisation of the looms and other coir manufacturing equipment are not widely accepted in this industry even if modern coir manufacturing machines are available.

Prevalence was determined with the percentage of positive responses to questions on musculoskeletal symptoms (Ghasemkhani 2008). Several papers reported that, the lower back pain (LBP) was the most prevalent body region. In a study among hospital staff in Turkey (n = 143), most of the participants (65.8%) had reported pain at lower back during the past 12 months, the highest prevalence of WMSDs reported was by nurses (77.1%) and the lowest amongst secretaries (54.1%) and hospital aides (53.5%) (Karahan et al. 2009). the lower back was the most common location of musculoskeletal pain among farmers (n = 3207) in Midwest states, USA (33.2%) (Fethke et al. 2015). Prevalence of WMSDs among the farmers of Kanpur-rural area in India, the most prevalent body regions were lower back (60%), knee (39%), shoulder (22%) and neck (10%) (Gupta 2013). In a study among the residents in Hong Kong reported that 21% (n = 652) had LBP in the past 12 months (Lau et al. 1995). In a study among physiotherapists (n = 3296) registered in the state of Victoria in Australia, it was reported that the most prevalence of WMSDs for the past 12 months was on the lower back (52.5%) then at neck 47.6% (Cromie et al. 2000). In a study among physiotherapists of Saurashtra region in India, the most prevalence of WMSDs reported at lower back (35%) followed by neck (25%) and shoulders (15%) (Buddhadev 2012). In a study among Iranian industrial workers (n = 10941), it was reported that 21% had LBP. The 12 months' prevalence in females was 7% higher than in males (Ghaffari et al. 2006). In a study among Iranian steel workers (n = 1439), LBP (64.12%) was the most commonly reported WMSDs (Aghilinejad et al. 2012). In a study among construction workers (n = 165) in Saudi Arabia, the majority of respondents had LBP (50%) followed by knee pain (20%) (Alghadir and Anwer 2015). In the following papers the most prevalence of WMSD reported other than lower back is discussed. In a study among clinical laboratory staff (n = 156) in Ethiopia, the major complaints were ankle/feet (21.7%) and knees (20.8%) (Haile et al. 2012). Among Brazilian hairdressers (n = 220), the most frequently affected body was the shoulder (49%), followed by the neck (47%) and back (39%) (Mussi and Gouveia 2008). In a study among dentists (n = 92) in Iran it was reported that the prevalent body regions were neck (43.4%), back (35.8%) and shoulder and wrist (each 25%) (Rabiei et al. 2015). In a study among dentists (n = 220) in Poland it was reported that the prevalent body regions were neck (47%), lower back (35%) and fingers (29%) (Kierklo et al. 2011). In a study among dental professionals (n = 533) in India, it was reported that the regions of symptoms were neck (75.74%), upper back (18.65%), lower back (72.01%), shoulder (69.4%), wrist (73.13%), hip (29.85%), elbow (7.46%) and ankle (12.31%) (Kumar et al. 2013).

During the literature survey, no papers were noticed regarding the prevalence study of WMSDs in coir industry workers. Some papers studied the textile and carpet weaving industry, which is similar to the coir mat and weaving industry. In a study among hand loom textile weavers (n = 175) of West Bengal state in India, reported that, the most prevalent WMSDs affected body part was the lower back (68%), followed by arm (49.7%), upper back (44%), knee (38%), shoulder (39.4%), wrist (35.4%) and neck (35.4%) (Durlov et al. 2014). In another study among textile weaving workers (n = 375) in state of Andhra Pradesh in India, reported that the most commonly affected body regions were lower back (60.6%), ankles/feet (59.0%), knee (58.1%) and upper back (54.6%). Female workers were significantly more affected than male workers in upper back, lower back, shoulders, wrists and ankles ($P < 0.05$) (Telaprolu and Anne 2014). In a study among Iranian hand woven carpet industry, workers (n = 1439), reported that the most commonly affected body regions were neck (35.2%), shoulders (47.8%), upper back (37.7%), lower back (45.2%), wrists (38.2%) and knees (34.6%) (Aireza Choobineh 2004). In a study among Nepalese textile workers, the 1-month prevalence of WMSDs, reported at the lower back was 35% (n = 324), being higher in females than males (45% vs. 28%; $P < 0.001$) (Paudyal et al. 2013).

10.2 Materials and Methods

A cross-sectional type of survey was conducted during the period of November 2015 to May 2016. The inclusion criteria were registered full-time coir workers from coir mat and matting industry. The 12 members of coir mat and matting industries are selected by convenient sampling method from the list of registered industries in Kerala. The self-reporting modified Nordic Musculoskeletal Questionnaires (NMQ) (Kuorinka et al. 1987) was distributed among the workers in person at the workplace during the working time and it was collected back within one week. 382 numbers of questionnaires were distributed to all the workers in coir mat and matting industries. Finally, 362 (response rate: 94.76%) questionnaires were collected from the returned questionnaires. The reason for mentioned response rate was rejection of incomplete and non-returned questionnaires. Along with the questionnaire a front page explaining the purpose of the study, general directions to fill up the questionnaire, contact mobile phone number were given and the second page contain a consent form for taking consent from the participant were attached. Workers were informed that their individual responses would remain anonymous and the data provided by them will be used only for this research work, and data will not be handed over to company management in any circumstances. The questionnaire consists of eight different parts (i) demographic (age, gender, etc.)—9 questions, (ii) health status—4 questions, (iii) workstation information—6 questions, (iv) working posture information—5 questions (v) Perceived pain and discomfort—3 questions, (vi) working time and schedules—6 questions (vii) physical condition in the factory—6 questions and (viii) workstation information—3 questions. The questionnaire was prepared in the mother tongue of the target population, Malayalam. Even though the validity and reliability of the NMQ were investigated in different studies, a thorough validation was conducted by a pilot study by circulating the questionnaire among 55 targeted workers and the questionnaire was modified on discussion with experts. Statistical analyses were conducted using the SPSS 20.0 package with a significance level set at $p < 0.05$ for all tests. The data collected was systematically entered, cleaned and analysed with the software. Pearson's chi-square analysis was applied to determine the association of prevalence of musculoskeletal symptoms with personal characteristics, duration of employment of workers and different task performed by the workers, etc. The chi-square statistic helps us to decide whether prevalence of WMSDs symptoms with different factors (personal characteristics, duration of employment, different task performed by the workers, etc.) were associated or not. It does not tell us how much more likely WMSDs would be expected to occur among workers, which the odds ratio does. In this study, logistic regression analysis was applied to find the Adjusted Odds Ratio (AOR) with 95% Confidence Interval, to quantify factors that were likely to affect WMSDs at different body regions. It is obvious that the odds ratio is a more interpretable statistic than the chi-square (Wang et al. 1995).

10.3 Results and Discussion

The demographics of the subjects are summarised in Table 10.1. The male and female workers participated in this study were found almost equal in numbers, male 184 (50.8%) and female 178 (49.2%), $n = 362$. This data was collected from coir mat and matting manufacturing industry only. But when considering the entire coir industry in Kerala the other reported studies indicate that the female population is about 70%. The percentage of workers whose age below 30 years was only 0.5%, age between 31 and 40 years was 6.6%, age between 41 and 50 was 34.2%, age between 51 and 60 was 35.4% and age above 60 years was 23.2%. It is very significantly higher than that of other studies. Only 0.5% workers are there in group below 30 years. Also, only 6.6% of workers are there in the group of age between 31 and 40 years. This indicates that the youngsters are not opting coir industry as a career opportunity, because of low income compared to the other jobs available in Kerala and also due to the fact that the physical work required is very heavy in this sector. In the case of educational qualification of workers, four female workers were fallen under the group no education (1.1%) which is very few in number. Most of the workers (45.6%) have primary (up to seventh standard) or secondary education (48.3%) (tenth standard). Workers beyond secondary education were comparatively less in number (5%). The association with educational qualification and WMSDs are not significant. In the case of monthly income of workers, majority of the workers (56.9%) are having monthly income below rupees 2500 ($37 approx.). The percentage of workers who are having monthly income between rupees 2500 and 5000 ($74 approx.), rupees 5000 and 7500 ($111 approx.) and above rupees 7500 were 32, 8.3, 2.8% respectively. It was noticed that most of the workers are having very low income and hence the workers are demotivated. The percentage of workers having duration of employment less than 10 years, between 11 years and 20 years, between 21 years and 30 years and beyond 31 years are 21.8, 40.1, 29.3 and 8.8% respectively. The Body Mass Index (BMI) of most of the workers (72.4%) falls under normal category. 13.6% workers fall under the category of underweight. 13.9% of workers fall under the category of overweight. It was noticed that most of the workers have normal BMI and no obese subjected are reported. From the Table 10.1, the workers having smoking habit is found to be nil in female workers and for the male workers it is only 11% of the total population, which is very small compared to the other studies and also there was no significant relation with WMSDs and smoking (p = 0.362). In a similar study at Meerut district in Utter Pradesh, India reported that the 80.1% of workers addicted to smoking or tobacco chewing habits (Sartaj 2013). Another study in Wardha district in Maharashtra state, India reported that, the 22.8% of the subjects were smokers (Tiwari et al. 2003).

The WMSDs in different body region are shown in Table 10.2. The 12-month prevalence rate of WMSDs was highest at lower back (58.6%). The disorder at lower back was higher for male (65.2%) workers than female (51.7%) workers. The region at knee (58.0%) was the second most prevalent WMSDs reported followed by elbow (42.8%), neck (39.8%), shoulder (37.6%), upper back (36.7%), ankle (25.4%), hip (15.7%) and wrist (15.5%). A few number of similar prevalence study of WMSDs among hand loom weavers were reported from various parts of India. Certain papers reported that most prevalent body region is lower back with a prevalent rate of 68, 60.6 and 68% respectively (Durlov et al. 2014; Pandit et al. 2013; Telaprolu and Anne 2014). This study is also agreeing with the said studies.

Table 10.1 Demographics of workers

Variable		Frequency		
		Female (n = 178)	Male (n = 184)	Total (%) (n = 362)
Age (Years)	Less than 30	2	0	2 (0.5%)
	Between 31 and 40	9	15	24 (6.6%)
	Between 41 and 50	73	51	124 (34.2%)
	Between 51 and 60	67	61	128 (35.4%)
	More than 60	27	57	84 (23.2%)
Level of Education	No Education	4	0	4 (1.1%)
	Primary	86	79	165 (45.6%)
	Secondary	80	95	175 (48.3%)
	Higher Sec./above	8	10	18 (5%)
Monthly Income (Rupees)	Less than 2500 ($37)	117	89	206 (56.9%)
	Between 2500 and 5000 ($74)	52	64	116 (32.0%)
	Between 5001 and 7500($ 111)	7	23	30 (8.3%)
	Above 7500 ($111)	2	8	10 (2.8%)
Duration of employment (Years)	Less than 10	50	29	79 (21.8%)
	Between 11 and 20	83	62	145 (40.1%)
	Between 21 and 30	43	63	106 (29.3%)
	Above 31	2	30	32 (8.8%)
Body Mass Index	Underweight, BMI $\leq$ 18.5	20	30	50 (13.8%)
	Normal, BMI = 18.5–24.9	133	129	262 (72.4%)
	Overweight, BMI = 25–29.9	25	25	50 (13.8%)
	Obesity BMI $\geq$ 30	0	0	0%
Section of Work	Winding (Spool and Babine)	142	8	150 (41.4%)
	Weaving	13	162	175 (48.3%)
	Warping	0	11	11 (3%)
	Spinning	23	3	26 (7.3%)
Smoking Habit	Smokers	0	40	40 (11%)

The second most prevalent body region in this study is knee and it is noticed that, in some of the study it was knee and in other studies it was at third or fourth position. In this study, high proportion of the population (90.3%) had experienced some form of WMSDs at one or other body region during the past 12 months. For female population, it was 90.2% and for male population it was 90.4%. Workers have no WMSDs symptoms affected was 9.7%. About 24% of workers have WMSDs

Table 10.2 Frequency of reported WMSD symptoms in different body regions of the male and female workers during the 12 months prior to the study

Body region	Female (n = 178)	Male (n = 184)	Total (n = 362)	Body region	Female (n = 178)	Male (n = 184)	Total (n = 362)
Neck	66 (31.7%)	78 (42.4%)	144 (39.8%)	Wrist	14 (7.9%)	42 (22.8%)	18 (15.5%)
Ur Back	62 (34.8%)	71 (38.6%)	133 (36.7%)	Hip	23 (13.9%)	34 (18.5%)	58 (15.7%)
Lr Back	92 (51.7%)	120 (65.2%)	212 (58.6%)	Knee	87 (48.9%)	122 (67.9%)	210 (58.0%)
Shoulder	57 (32.0%)	79 (42.9%)	136 (37.6%)	Ankle	38 (21.3%)	54 (29.3%)	92 (25.4%)
Elbow	56 (31.5%)	99 (53.8%)	155 (42.8%)				

Number of body region affected the WMSD	No of workers reported (%)		Total (%) (n = 362)
	Male (n = 184)	Female (178)	
No WMSD	17 (4.7%)	18 (5.0%)	35 (9.7%)
1	65 (18.0%)	22 (6.1%)	87 (24.0%)
2	31 (8.6%)	35 (9.7%)	66 (18.2%)
3	18 (5.0%)	21 (5.8%)	39 (10.8%)
4	15 (4.1%)	29 (8.0%)	44 (12.2%)
5 or more	32 (8.8%)	59 (16.3%)	91 (25.1%)
Total	161 (90.4%)	166 (90.2%)	327 (90.3%)

symptoms at any one body region. WMSDs affected at more than one body region are given in the Table 10.3.

From the Table 10.4, it is shown that the prevalence of WMSDs at the body region neck (p = 0.013), upper back (p = 0.048), lower back (p = 0.007), shoulder (p = 0.006), elbow (p = 0.001), wrist (p = 0.001), hip (p = 0.006), knee (p = 0.001) and ankle (p = 0.001) was significantly associated with gender. The WMSDs at neck, upper back, lower back, shoulder, elbow, wrist, hip, knee and

Table 10.3 Genderwise Distribution of Number of Body Regions affected by WMSDs

Body Regions affected	Percent of persons affected in		Proportion of total number of persons affected (in percent)
	Male	Female	
No effect	4.7	5.0	9.7
In one only	18.0	6.1	24.0
In two only	8.6	9.7	18.2
In three only	5.0	5.8	10.8
In four only	4.1	8.0	12.2
In five or more	8.8	16.3	25.1

Table 10.4 Association of WMSD symptoms at different body region with gender

Body region	*P** Value	OR (at 95% CI)	Body region	*P** Value	OR (at 95% CI)
Neck	0.013**	2.514 (1.214–5.206)	Wrist	0.001**	11.887 (4.128–33.789)
Upper Back	0.048**	1.621 (0.961–2.736)	Hip	0.006**	3.755 (1.470–9.595)
Lower Back	0.007**	1.805 (1.175–2.772)	Knee	0.001**	2.963 (1.833–4.789)
Shoulder	0.006**	2.690 (1.337–5.413)	Ankle	0.001**	3.729 (1.673–8.314)
Elbow	0.001**	4.588 (2.238–9.405)			

*χ^2 analysis of prevalence of MSD symptoms at different body region with gender. **Statistically significant with $p < 0.05$, OR Odds Ratio with female as reference category

ankle of male workers was 2.514 (OR), 1.621, 1.805, 2.690, 4.588, 11.887, 3.755, 2.963 and 3.729 times higher that of female workers. The WMSDs for male were significantly more affected than female in all the body regions. In a study of gender differences in WMSDs of a similar industry like textile sector reported that, the female weavers in power loom and hand loom were more prone to developing WMSDs in upper back and lower back and male weavers were more prone to developing pain in the knee and hand (Nag et al. 2010).

In this study workers are grouped in four groups based on the age. Since number of workers with age less than 30 years was only 2 which is very less when compared with other groups hence added with next group. Workers with age less than 40 years are considered as first group (group A). The second group was age between 41 and 50 (group B). The third group was age between 51 and 60 (group C). The last group was age above 60 years (group D). The prevalence of WMSDs at the body regions of neck ($p = 0.018$), shoulder ($p = 0.001$), elbow ($p = 0.003$), hip ($p = 0.015$), knee ($p = 0.004$) and ankle ($p = 0.038$) was significantly associated with age group. The WMSDs at neck for the workers having age between 41 and 50 was 1.274 (OR) times higher symptoms reported than workers age less than 40 years, similarly for group C and D compared to group A, OR was 2.347 and 1.002 times higher and for the age group C and D compared to group B, OR was 1.252 and 1.525 times higher WMSD reported. For the group D compared to group C it was 1.250. Similarly, for the body region shoulder, elbow, hip, knee and ankle were given in the Table 10.5. It may be concluded that, when the age of the workers increases the WMSDs symptoms also found to be increasing at the regions of shoulder, elbow, hip, knee and angle.

In this study, the duration of employment of the workers was categorised into four groups. The prevalence of WMSDs at the body region neck ($p = 0.032$), shoulder ($p = 0.031$), elbow ($p = 0.036$), wrist ($p = 0.002$) hip ($p = 0.015$) and knee ($p = 0.004$) was significantly associated with different categories of duration of employment. The WMSDs at neck of workers having employment duration

Table 10.5 Association of WMSD symptoms at different body region with age groups

Body region	*P** Value	Adjusted OR (at 95% Confidence Interval)					
		A–B#	A–C#	A–D#	B–C#	B–D#	C–D#
Neck	0.018**	1.274 (0.453–3.597)	2.347 (1.176–4.675)	1.002 (0.498–1.976)	1.252 (0.578–2.154)	1.525 (0856–2.101)	1.250 (0.684–2.524)
Upper Back	0.443†	–	–	–	–	–	–
Lower Back	0.720†	–	–	–	–	–	–
Shoulder	0.001**	0.816 (0.324–2.054)	1.442 (0.583–3.567)	2.596 (1.017–6.624)	1.769 (1.037–3.014)	3.182 (1.772–5.713)	1.800 (1.032–3.141)
Elbow	0.003**	0.805 (0.330–1.962)	1.368 (0.604–3.525)	2.032 (0.962–6.396)	1.889 (1.125–3.171)	3.001 (1.648–5.464)	1.486 (0.854–2.584)
Wrist	0.666†	–	–	–	–	–	–
Hip	0.015**	1.747 (0.360–8.480)	3.044 (0.651–14.22)	5.811 (1.174–28.74)	1.826 (0.853–3.908)	3.364 (1.438–7.871)	1.656 (0.776–3.537)
Knee	0.004**	0.518 (0.212–1.265)	0.915 (0.371–2.258)	1.595 (0596–4.271)	1.720 (1.032–2.865)	3.130 (1.674–5.852)	1.833 (1.002–3.352)
Ankle	0.038**	0.366 (0.141–0.954)	0.703 (0.276–1.790)	0.881 (0.333–2.033)	2.033 (1.098–3.767)	2.2294 (1.175–4.481)	1.175 (0.640–2.154)

*χ^2 analysis of prevalence of WMSD symptoms between age groups. **Statistically significant with $p < 0.05$, †Not significant. #*A* Age group less than 40 years, *B* Age group in between 41 and 50 years, *C* Age group in between 51 and 60 years and *D* Age group above 60 years

between 11 and 20 years are 1.146 (OR) times higher than workers having less than 10 years of duration of employment. The WMSDs at neck of workers in the group C and group D were 0.813 and 0.277 (OR) times less than workers in the group A respectively. The WMSDs at neck of workers in the group C and group D were 0.696 and 0.222 (OR) times less than workers in the group B, respectively. The WMSDs at neck of workers in the group D were 0.370 (OR) times less than workers in the group C. The WMSDs at shoulder, elbow, wrist and hip were also higher for weaving and warping workers than spinning workers. The Table 10.6 gives the value of adjusted OR in each case. From the Table 10.6 it may concluded that, when the duration of employment increases, the symptoms of WMSDs were also found to be increasing at the body region, shoulder, elbow, wrist, hip and knee, but it was found to be decreasing at the neck.

In this study, the body mass index (BMI) of the workers was categorised to three groups, viz. normal, underweight and overweight. None of the workers fall under the group obese. The prevalence of WMSDs at the body region neck ($p = 0.024$), lower back ($p = 0.046$), shoulder ($p = 0.041$), wrist ($p = 0.001$), hip ($p = 0.003$),

Table 10.6 Association of WMSD symptoms at different body regions with duration of employment

Body region	P* Value	Adjusted OR (at 95% Confidence Interval)					
		A–B#	A–C#	A–D#	B–C#	B–D#	C–D#
Neck	0.032**	1.146 (0623–2.107)	0.813 (0429–1.542)	0.277 (0.100–0.277)	0.696 (0.409–1.186)	0.226 (0.081–0.632)	0.370 (0.142–0.964)
Upper Back	0.262†	–	–	–	–	–	–
Lower Back	0.658†	–	–	–	–	–	–
Shoulder	0.031**	1.366 (0.749–2.493)	2.421 (1.297–4.517)	2.751 (1.738–4.154)	1.612 (0.941–2.763)	1.241 (0.531–2.898)	0.730 (0.327–1.633)
Elbow	0.036**	1.370 (0.757–2.480)	1.404 (1.086–4.493)	1.487 (0.617–3.579)	1.586 (0.936–2.699)	0.948 (0.423–2.126)	0.571 (0.252–1.294)
wrist	0.002**	7.258 (1.664–31.66)	11.268 (2.576–49.28)	11.78 (2.102–55.29)	1.245 (0.629–2.499)	1.340 (0.496–3.623)	0.994 (0.375–2.632)
Hip	0.013**	0.736 (0.316–1.713)	2.319 (1.040–5.170)	1.349 (0.415–4.380)	2.822 (1.374–5.795)	1.604 (0.524–4.911)	0.613 (0.212–1.769)
Knee	0.010**	0.282 (0.264–0.80)	0.321 (0.164–0.626)	0.39 (0.153–1.032)	0.655 (0.386–1.145)	0.824 (0.348–1.948)	1.239 (0.527–2.912)
Ankle	0.698†	–	–	–	–	–	–

*χ^2 analysis of prevalence of WMSD symptoms at different body region against different categories of duration of employment. **Statistically significant with $p < 0.05$, †Not significant. #*A* duration of employment less than 10 years, *B* between 11 and 20 years, *C* between 21 and 30 years, *D* more than 31 years

knee ($p = 0.029$), and ankle ($p = 0.010$) was significantly associated with different categories of BMI. The WMSDs at neck of workers within the underweight category and overweight category were 1.110 (OR) and 2.709 times higher than workers having normal BMI. The WMSDs at lower back, shoulder, wrist, hip, knee and ankle were also higher for weaving and warping workers than spinning workers. Table 10.7 gives the value of adjusted OR in each case.

In this study, four major tasks are considered and they are spinning, winding, weaving and warping. The prevalence of WMSDs at the body region upper back ($p = 0.016$), lower back ($p = 0.041$), shoulder ($p = 0.041$), elbow ($p = 0.003$), wrist ($p = 0.001$), knee ($p = 0.001$) and ankle ($p = 0.024$) except neck was significantly associated with different task. WMSDs at lower back of weaving, warping and spinning workers were 1.039, 4.203, and 4.061 (OR) times higher than winding workers respectively. Table 10.8 gives the value of adjusted OR at different body regions. The WMSDs symptoms for the weaving workers were higher

Table 10.7 Association of WMDS symptoms at different body region with BMI groups

Body region	*P** Value	Adjusted OR(at 95% Confidence Interval)	
		A–B#	A–C#
Neck	0.024**	1.110 (0.602–2.045)	2.709 (1.179–6.223)
Upper Back	0.254†	–	–
Lower Back	0.046**	2.330 (1.192–4.556)	2.186 (0.889–5.376)
Shoulder	0.041**	1.833 (0.934–3.596)	3.123 (1.287–7.576)
Elbow	0.246†	–	–
Wrist	0.001**	2.020 (0.689–5.924)	6.944 (2.095–23.014)
Hip	0.003**	2.148 (0.734–6.284)	6.250 (1.876–20.827)
Knee	0.029**	2.289(1.222–4.289)	2.445 (1.015–5.892)
Ankle	0.010**	1.041 (0.516–2.102)	2.891 (1.195–6.996)

*χ^2 analysis of prevalence of MSD symptoms at different body region with BMI groups. **Statistically significant with $p < 0.05$, †Not significant. #*A* Normal Weight (BMI = 18.5–24.9), *B* Under Weight (BMI < 18.5), *C* Over weight (BMI = 25–29.9)

Table 10.8 Association of WMSD symptoms at different body region with different section of work

Body region	*P** Value	Adjusted OR (at 95% Confidence Interval)					
		A–B#	A–C#	A–D#	B–C#	B–D#	C–D#
Neck	0.211†	–	–	–	–	–	–
Upper Back	0.016**	1.039 (0.612–0.765)	4.203 (11.03–6.00)	4.061 (1.427–1.55)	2.875 (0.778–0.629)	2.462 (0.971–0.245)	1.124 (0.313–0.782)
Lower Back	0.041**	1.773 (1.088–0.759)	5.416 (1.055–7.52)	1.110 (0.474–0.601)	3.097 (0.636–0.078)	0.522 (0.209–0.304)	0.235 (0.052–0.050)
Shoulder	0.041**	0.134 (0.034–0.530)	0.166 (0.026–0.076)	0.829 (0.312–0.204)	1.304 (0.342–0.968)	5.661 (1.24–25.78)	3.887 (0.74–20.42)
Elbow	0.003**	2.326 (1.476–0.665)	1.228 (0.343–0.400)	1.842 (0.791–0.289)	0.528 (0.149–0.868)	0.792 (0.347–0.808)	1.429 (0.400–0.099)
Wrist	0.001**	2.984 (1.495–0.959)	0.728 (0.088–0.996)	2.599 (0.831–0.131)	0.244 (0.031–0.909)	0.871 (0.308–0.462)	2.352 (0.241–0.937)
Hip	0.212†	–	–	–	–	–	–
Knee	0.001**	2.161 (1.375–0.395)	1.750 (0.491–0.232)	0.625 (0266–1.467)	0.810 (0.228–0.880)	0.289 (0.123–0.678)	0.198 (0.049–0.804)
Ankle	0.024**	1.549 (0.923–0.594)	0.393 (0.048–0.194)	2.884 (1.202–0.921)	0.254 (0.032–0.036)	1.863 (0.801–0.332)	5.589 (2.125–0.254)

*χ^2 analysis of prevalence of MSD symptoms at different body region with different section of work. **Statistically significant with $p < 0.05$, †Not significant. #*A* Winding, *B* Weaving, *C* Warping, *D* Spinning

Table 10.9 Association of WMDs symptoms at different body region with education

Body region	P^* Value	Comment (at 95% CI)	Body region	P^* Value	Comment (at 95% CI)
Neck	0.297	Not Significant	Wrist	0.650	Not Significant
Upper Back	0.243	Not Significant	Hip	0.370	Not Significant
Lower Back	0.182	Not Significant	Knee	0.814	Not Significant
Shoulder	0.717	Not Significant	Ankle	0.418	Not Significant
Elbow	0.311	Not Significant			

$^*\chi^2$ analysis of prevalence of MSD symptoms at different body regions against education

Table 10.10 Association of WMSDs at any of the body region with the physical condition

Body region	P^* Value	Comment (at 95% CI)	Body region	P^* Value	Comment (at 95% CI)
Light	0.158	Not Significant	Dust	0.021	Significant
Heat	0.001	Significant	Odour	0.019	Significant
Noise	0.002	Significant			

$^*\chi^2$ analysis of prevalence of WMSD symptoms at different body region with physical condition

at the body regions of upper back, lower back, elbow, wrist, knee and ankle. The symptoms were lower at shoulder and hip than winding workers. WMSDs symptoms for the warping workers were higher at the body regions of upper back, lower back elbow, wrist, hip and knee and the symptoms were lower at shoulder and ankle than winding workers. The WMSDs symptoms for the spinning workers were higher at the body region of upper back, lower back elbow, wrist, hip and ankle and the symptoms were lower at shoulder and knee than winding workers.

The prevalence of WMSDs was not significantly associated with the education of workers (Table 10.9). It is also identified that the significant relationships were existing with physical factors such as heat ($p = 0.001$), noise ($p = 0.002$), dust ($p = 0.021$) and odour ($p = 0.019$) except light ($p = 0.152$) at 95% CI (Table 10.10).

10.4 Conclusions

WMSDs are an important health risk among the coir industry workers. This study provides data on the prevalence of WMSDs in the coir industry workers in the state of Kerala, India. The present investigation showed that there was a high rate of WMSDs problems among the workers in all sections of coir mat and matting industry. Therefore, it can be concluded that to improve the working conditions and to reduce the WMSDs problems in this industry we should be focused on designing of workstations with the application of ergonomic principles. It could be confirmed that it would significantly enhance the productivity and working efficiency of workers to a great extent.

Acknowledgements The authors express sincere gratitude to the managements and workers of the coir industries for their cooperation and support for the completion of this study.

References

Aghilinejad M, Choobineh AR, Sadeghi Z, Nouri MK (2012) Prevalence of musculoskeletal disorders among Iranian steel workers. Iran Red Crescent Med J, 198–203

Aireza Choobineh ML (2004) Musculoskeletal symptoms as related to ergonomic factors in Iranian hand-woven carpet industry and general guidelines for workstation design. Int J Occup Saf Ergon (JOSE) 10(2):157–168

Alghadir A, Anwer S (2015) Prevalence of musculoskeletal pain in construction workers in Saudi Arabia. Sci World J 2015:1–5

Australia Safety Council (2006) Work-related musculoskeletal disease in Australia. Commonwealth of Australia, Canberra, pp 326–770

Buddhadev Neeti P (2012) Work related musculoskelletal disoders: a survey of physiotherapists in Sourashtra region. Natl J Med Sci 2(2):179–181

Cromie JE, Robertson VJ, Best MO (2000) Work-related musculoskeletal disorders in physical therapists: prevalence, severity, risks, and responses. Phys Ther 80(4):336–351

Durlov S, Chakrabarty S, Chatterjee A, Das T, Dev S, Gangopadhyay S, Haldar P, Maity SG, Sarkar K, Sahu S (2014) Prevalence of low back pain among handloom weavers in West Bengal, India. Int J Occup Environ Health 20(4):333–339

Fethke NB, Merlino LA, Gerr F, Schall MC, Branch CA (2015) Musculoskeletal pain among Midwest farmers and associations with agricultural activities. Am J Ind Med 58(3):319–330

Ghaffari M, Alipour A, Jensen I, Farshad AA, Vingard E (2006) Low back pain among Iranian industrial workers. Occup Med 56(7):455–460

Ghasemkhani M (2008) Musculoskeletal symptoms in workers. Int J Occup Saf Ergon

Gupta G (2013) Prevalence of musculoskeletal disorders in farmers of Kanpur-Rural, India. J Community Med Health Educ 3(7)

Haile EL, Taye B, Hussen F (2012) Ergonomic workstations and work-related musculoskeletal disorders in the clinical laboratory. Lab Med 43(suppl 2):11–19

Karahan A, Kav S, Abbasoglu A, Dogan N (2009) Low back pain: prevalence and associated risk factors among hospital staff. J Adv Nurs 65(3):516–524

Kerala Economic Review (2015) Chapter 3, Industry labour and employment. State Planning Board Kerala, Thiruvananthapuram, India

Kierklo A, Kobus A, Jaworska M, Botuliñski B (2011) Work-related musculoskeletal disorders among dentists-a questionnaire survey. Ann Agric Environ Med 18(1)

Kumar VK, Kumar SP, Baliga MR (2013) Prevalence of work-related musculoskeletal complaints among dentists in India: a national cross-sectional survey. Ind J Dent Res 24(4):428

Kuorinka I, Jonsson B, Kilbom A, Vinterberg H, Biering-Sørensen F, Andersson G, Jørgensen K (1987) Standardised Nordic questionnaires for the analysis of musculoskeletal symptoms. Appl Ergon 18(3):233–237

Lau EM, Egger P, Coggon D, Cooper C, Valenti L, O'Connell D (1995) Low back pain in Hong Kong: prevalence and characteristics compared with Britain. J Epidemiol Community Health 49(5):492–494

Mussi G, Gouveia N (2008) Prevalence of work-related musculoskeletal disorders in Brazilian hairdressers. Occup Med 58(5):367–369

Nag A, Vyas H, Nag PK (2010) Gender differences, work stressors and musculoskeletal disorders in weaving industries. Ind Health 48(3):339–348

Nunes IL, Bush PM (2012) Work-related musculoskeletal disorders assessment and prevention. In: Ergonomics—a systems approach. INTECH Open Access Publisher, pp 1–31

Pandit S, Kumar P, Chakrabarti D (2013) Ergonomics problems prevalent in handloom unit of North East India. Int J Sci Res Publ 3(1):1–7

Paudyal P, Ayres JG, Semple S, Macfarlane GJ (2013) Low back pain among textile workers: a cross-sectional study. Occup Med 63(2):129–134

Podniece Z, Heuvel S, Blatter B (2008) Work-related musculoskeletal disorders: prevention report. European Agency for Safety and Health at Work, Office for Official Publications of the European Communities

Putz-Anderson V, Bernard BP, Burt SE, Cole LL, Fairfield-Estill C, Fine LJ, Grant KA, Gjessing C, Jenkins L, Hurrell JJ Jr, Nelson N (1997) Musculoskeletal disorders and workplace factors. National Institute for Occupational Safety and Health (NIOSH)

Rabiei M, Shakiba M, Dehgan Shahreza H, Talebzadeh M (2015) Musculoskeletal disorders in dentists. Int J Occup Hyg 4(1):36–40

Sartaj A (2013) Musculoskeletal disorders and respiratory illness of workers in small scale textile industries in Meerut district. Asian J Pharm Hea Sci 3(4):882–886

Telaprolu N, Anne SD (2014) Physical and psychological work demands as potential risk factors for musculoskeletal disorders among workers in weaving operations. Ind J Occup Environ Med 18(3):129

Tiwari RR, Pathak MC, Zodpey SP (2003) Low back pain among textile workers. Ind J Occup Environ Med 7(1):27–29

Wang MQ, Eddy JM, Fitzhugh EC (1995) Application of odds ratios and logistic models in epidemiology and health research. Health Values 19(1):59–62

Chapter 11
Musculoskeletal Disorders and Workplace Analysis of Assembly Section in a Submersible Pump Industry: An EMG-Based Approach

P. Ashok, G. Madhan Mohan and K. Nitish

Abstract In manufacturing industries, poor working posture is a common issue that may lead to work-related Musculoskeletal Disorder (WMSD) and physiological stress among workers. Poor working postures are considered as the major contributor that affects the worker efficiency and productivity. The purpose of the study is to enumerate the causes that contribute to musculoskeletal disorder among the workers. Thirty-nine male workers with in an age group of 20–24 years participated in the study. Responses as obtained through administration of Musculoskeletal Questionnaire, are analyzed by Rapid Upper Limb Analysis (RULA) method, Heart Rate (HR) monitoring and Surface Electromyography (sEMG). The heart rate was continuously monitored by using heart rate sensor for 4 h and concurrently the muscle activity has been monitored by sEMG. The results from the Nordic questionnaire and sEMG are similar. It is observed that more than 58 percent of the workers affected by MSD and physiological stresses due to awkward working postures and poor workplace design. From this study, it can be concluded that the efficiency and productivity of workers can be improved by modifying working postures and workplace.

Keywords RULA · Heart rate monitor · Electromyography · Physiological stress · Working postures

11.1 Introduction

Musculoskeletal Disorder (MSD) is the major work-related problem in any industry particular in manual material and assembly industries (Pinto et al. 2013). Physiological stresses are very much related to oxygen consumption, increased heart rate and workplace design (Whitfield et al. 2014). The effect of grip force and

P. Ashok (✉) · G. Madhan Mohan · K. Nitish
Department of Production Engineering, PSG College of Technology, Coimbatore, Tamil Nadu, India
e-mail: ashokdino@gmail.com; pak@pro.psgtech.ac.in

P.K. Ray and J. Maiti (eds.), *Ergonomic Design of Products and Worksystems—21st Century Perspectives of Asia*, Managing the Asian Century,
https://doi.org/10.1007/978-981-10-5457-0_11

working postures are responsible for work-related problems in many cases (Finneran and O'Sullivan 2013). The effect of pulling and lifting tasks will also increase the level of MSD (Basahel 2015). For prolonged standing or in a particular posture for a long time causes pain and thus the efficiency of the worker gets reduced (Bjoringa and Hagga 2000). For an assembly section, it is important that the things needed for the assembly should be placed nearby to the worker for complete the task with less effort (Sullivan and Gallwey 2002). When the work load of the worker increases, the cardiovascular system responds and the heart rate as well as blood pressure will increase as a result the worker gets tired faster (Fredericksa et al. 2005). Workers having neck or shoulder complaint will affect more during difficult work (Kallenberg et al. 2006). When the load on the trapezius muscle increases the risk in neck and shoulder will increases (Aaras 1994). Lower back and musculoskeletal disorders of workers can be reduced by proper workplace design or by using personal assistance devices (Paskiewicza and Fathallaha 2007). Work impairment and work ability is reduced those who having musculoskeletal disorders (Reeuwijk et al. 2014). Prevention of work-related MSDs is beneficial for workers, employers and society by implementing job rotation (Comper and Padula 2014). Functional capacity evaluation (FCE) is one of the complementary value to assess the MSD among sitting physicians (Wind et al. 2009). WMSDs rate are high among women and young workers (Kang et al. 2014). Work intensification increased WMSDs, which created massive workers compensation collective claims. Workers argued for the reduction of work intensity (Kang et al. 2014). The proper implementation of ergonomic workplace will reduce the work-related disorders and the productivity of the organization can be improved (Gangopadhyay et al. 2014). Work-related disorders are similar for most of the industries as well as regions (Kim and Nakata 2014). It is one of the functions of the industry to ensure the comfort of each workers, since workers are the assets of each industry (Boatcaa and Cirjaliub 2015). Battini et al. (2015) conducted a study using two methods. The first one is multiobjective model based on energy expenditure that is used to estimate ergonomic level and second one transforms the energy expenditure rate during rest time. There should be a good provision for medical and psychological care for the injured workers for their speedy recovery of work-related disorders (Choi et al. 2014).

11.2 Methodology

11.2.1 Subjects

Thirty-nine male healthy subjects were selected for this study (Table 11.1). Mean and standard deviation (SD) anthropometric data were: 22.4 years of age, 172.5 cm in height and 68.0 kg in mass.

Fig. 11.1 Worker with awkward posture

11.2.2 Description of Tasks and Industry

This study investigated the postures specifically in lower house assembly of a motor in a submersible pump. The current study aimed to identify the ergonomics analysis, in particular evaluation and methods to reduce that can be associated with both these tasks. The assembly tasks were selected because workers have poor working postures (Fig. 11.1), manual material handling and impact loads.

The industry produces large number of submersible pumps per day. The industry includes around 12 sections of assembly with different layouts, and the major sections depend on manual activities and handling with poor types of manual tools and equipment with poor working postures. The work shift is from 8:30 am to 5:00 pm, 48 h/week in this type of tasks. The workers are given two break periods and one hour for lunch.

The various pump casing by hammering and whole assembly weighing from 16 to 25 kg. Before and after assembly, the worker needs to carry the parts form ground to the table and vice versa. For a particular task of lower house assembly, the worker needs to be in a position of bending down and hammering the part with high force which causes an impact load to the arm. Due to poor working posture, pain occurred in the arm, neck and back of the worker. The problems were occurred due to poor workplace design.

11.2.3 Experimental Procedure

The experiment has been started by monitoring work postures of each worker. The difficulties of the worker has been recorded by interviews and by using Nordic questionnaire. From Borg scale, the nature of the work can be recorded clearly.

Table 11.1 Anthropometric data of workers

Body dimension	50th percentile, cm	Range, cm min-max
Stature height	172.6	157.0–181.0
Eye height	161.3	144.0–168.0
Shoulder height	145.0	130.0–150.0
Elbow height	109	96.0–115.0
Knuckle height	75.3	70.2–80.4
Sitting height erect	91.3	85.1–97.5
Eye height sitting	79.3	73.4–85.2
Elbow rest height	27.8	25.4–30.2
Thigh clearance height	12.9	11.0–15.2
Knee height	56.4	52.0–60.0
Popliteal height	42.2	40.0–44.0
Buttock–knee length	57.7	55.0–60.0
Hip breadth	32	27.5–36.0
Weight	72.3	60.0–85.0

Table 11.2 Muscle names

Area	Name of the muscle
Upper arm	Brachioradialis
Behind the neck	Trapezius
Back	Thoracolumbar fascia

The preliminary analysis has been carried out by obtaining Rapid Upper Limb Analysis (RULA) score. The awkward postures from the industry has been modelled and analyzed in CATIA software. The heart rate variations of the worker have been monitored continuously by using heart rate sensor.

After the preliminary analysis, the detailed muscle activity is measured by surface Electromyography (sEMG) technique for finding the activity of Brachioradialis, Trapezius and Thoracolumbar fascia muscles. As per the data collected from questionnaire and interviews, the above mentioned muscles have been selected because of the pain among workers at the time of assembly. The subjects having excess hair shaved and the area has been rubbed with alcohol assure proper electrode–skin contact. Before and in between the measurement, the signal quality was inspected visually. When excessive noise was present, a small amount of gel was added to the electrodes or they were repositioned until signals with good quality were obtained Table 11.2.

The sensors have been carefully placed on the trapezius and thoracolumbar muscles after each shift of work. The experiments have been carried out continuously in existing workplace (Fig. 11.2) having poor working postures and modified workplace (Fig. 11.3). The results from the experiments have been compared and the modified workplace can be validated. The modified workplace is designed in such a way that the worker can assemble parts to the pump according to their comfortable height and posture.

Table 11.3 EMG data for different muscles during present and modified workplaces

Workplace	Muscle	Shift	Condition	Normalized EMG (%)
Present	Brachioradialis	Shift 1	Rest	32.7
			Working	74.6
		Shift 2	Rest	38.3
			Working	78.7
Modified	Brachioradialis	Shift 1	Rest	33.8
			Working	72.3
		Shift 2	Rest	36.8
			Working	76.4
Present	Trapezins	Shift 1	Rest	34.7
			Working	62.7
		Shift 2	Rest	37.2
			Working	64.3
Modified	Trapezins	Shift 1	Rest	28.3
			Working	48.9
		Shift 2	Rest	33.7
			Working	52.3
Present	Thoracolumbar Fascia	Shift 1	Rest	43.8
			Working	79.8
		Shift 2	Rest	46.4
			Working	83.7
Modified	Thoracolumbar Fascia	Shift 1	Rest	37.6
			Working	54.9
		Shift 2	Rest	31.2
			Working	53.4

Fig. 11.2 Present workplace

Fig. 11.3 Modified workplace

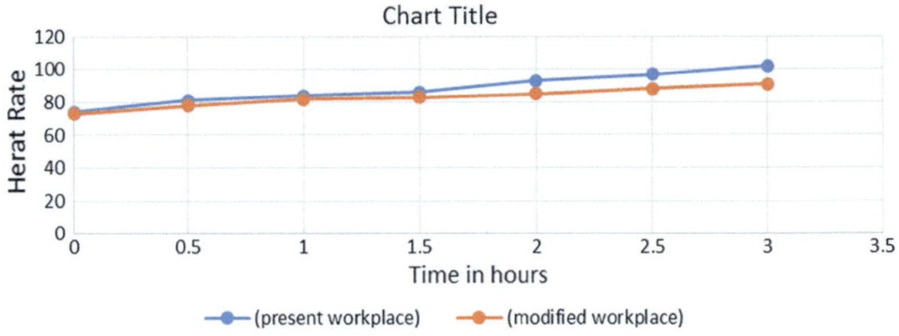

Fig. 11.4 Heart rate monitoring graph

11.3 Results

11.3.1 Heart Rate (HR)

The heart rate sensor is generally used to monitor the heart beat variation continuously to evaluate the job nature and working environment. In this study, the rate has been monitored every half an hour and the values were recorded. Here, the analysis shows that there is an increase in heart rate of the worker after each shift. The heart rate has been monitored at present workplace as well as modified workplace. Figure 11.4, shows that for a modified workplace, the variation of heart rate is lesser than that of present workplace.

Table 11.4 RULA Scores

Workplace	Upper arm	Neck	Trunk	Leg	Final RULA Score
Existing	4	6	7	5	6
New	3	2	2	3	3

11.3.2 RULA Method

RULA scores showed that the present workplace is awkward as compared to modified one.

Table 11.4 shows that neck, trunk, leg and upper arm is the major contribution of RULA score which determines the comfortless of the working postures.

11.3.3 Electromyography Analysis

The activity of Brachioradialis, Trapezins and Thoracolumbar fascia muscles has been evaluated by EMG technique. A summary of test information is shown in Table 11.3. The data shows that the normalized EMG during working and rest conditions for two shifts. The data has been recorded for the two workplaces for comparison. The EMG reading has been normalized for the purpose of comparison. During working condition, the activity of Brachioradialis muscle, for modified workplace (72.3) is lower than present (74.6). In the case of Trapezins muscle for modified workplace (48.9) is lower than present (62.7). Also, the Thoracolumbar fascia muscle for modified workplace (54.9) is lower than present (79.8). The above results show that for modified workplace, the work-related musculoskeletal disorder (WMSD) can be reduced.

11.4 Conclusions

The musculoskeletal disorder and the physiological stress cannot be fully avoided in an industry but can be reduced by suitable working environment and worker friendly workplace design. From the study, it is proved that workplace design is one of the major factor for musculoskeletal disorder and other work-related problems. The industry workplace should be designed as per worker comfort in order to reduce the musculoskeletal disorders and other work-related problems and there by the productivity of the industry can be improved.

References

Aaras A (1994). Relationship between trapezius load and the incidence of musculoskeletal illness in the neck and shoulder. Int J Ind Ergon 14: 341–348

Basahel Abdulrahman M (2015) Investigation of work-related Musculoskeletal Disorders (MSDs) in warehouse workers in Saudi Arabia. Procedia Manufacturing 3:4643–4649

Battini D, Delorme X, Dolgui A, Sgarbossa F (2015) Assembly line balancing with ergonomics paradigms: two alternative methods. IFAC-Papers OnLine 48–3:586–591

Bjoringa G, Hagga GM (2000) Musculoskeletal exposure of manual spray painting in the woodworking industry—an ergonomic study on painters. Int J Ind Ergon 26:603–614

Boatcaa ME, Cirjaliub B (2015) A proposed approach for an efficient ergonomics intervention in organizations. Procedia Economics and Finance 23:54–62

Choi M, Kim HR, Lee J (2014) Workers' experiences with compensated sick leave due to musculoskeletal disorder: a qualitative study Annals of. Occup Environ Med 26:33

Comper MLC, Padula RS (2014) The effectiveness of job rotation to prevent work-related musculoskeletal disorders: protocol of a cluster randomized clinical trial. Comper and Padula BMC Musculoskeletal Disorders 15:170

Eun-A Kim and Minori Nakata (2014) Work-related musculoskeletal disorders in Korea and Japan: a comparative description. Ann Occup Environ Med 26:17

Finneran Aoife, O'Sullivan Leonard (2013) Effects of grip type and wrist posture on forearm EMG activity, endurance time and movement accuracy. Int J Ind Ergon 43:91–99

Fredericksa Tycho K, Choib Sang D, Harta Jason, Butta Steven E, Mital Anil (2005) An investigation of myocardial aerobic capacity as a measure of both physical and cognitive workloads. Int J Ind Ergon 35:1097–1107

Gangopadhyay S, Dev S, (2014) Design and evaluation of ergonomic interventions for the prevention of musculoskeletal disorders in India. Ann Occup Environ Med 26:18

Kallenberg Laura AC, Hermens Hermie J, Vollenbroek-Hutten Miriam MR (2006) Distinction between computer workers with and without work-related neck—shoulder complaints based on multiple surface EMG parameters. Int J Ind Ergon 36:921–929

Kang Dongmug, Kim Young-Ki, Kim Eun-A, Kim Dae Hwan, Kim Inah, Kim Hyoung-Ryoul, Min Kyoung-Bok, Jung-Choi Kyunghee, Sung-Soo Oh, Koh Sang-Baek (2014a) Prevention of work-related musculoskeletal disorders. Ann Occup Environ Med 26:14

Kang Dongmug, Kim Youngki, Lee Young-Il, Koh Sangbaek, Kim I, Lee H (2014b) Work-related musculoskeletal disorders in Korea provoked by workers' collective compensation claims against work intensification. Ann Occup Environ Med 26:19

Paskiewicza Jessica K, Fathallaha Fadi A (2007) Effectiveness of a manual furniture handling device in reducing low back disorders risk factors. Int J Ind Ergon 37:93–102

Pinto VJ, Sheikhzadeh A, Halpern M, Nordin M (2013). Assessment of engineering controls designed for handling unstable loads: an electromyography assessment. Int J Ind Ergon 43:181–186

Reeuwijk Kerstin G, Robroek Suzan J W, Hakkaart Leona, Burdorf Alex (2014) How work impairments and reduced work ability are associated with health care use in workers with musculoskeletal disorders, cardiovascular disorders or mental disorders. J Occup Rehabil 24:631–639

Sullivan LWO, Gallwey TJ (2002) Effects of gender and reach distance on risks of musculoskeletal injuries in an assembly task. Int J Ind Ergon 29:61–71

Whitfield Brett H, Costigan Patrick A, Stevenson Joan M, Smallman Catherine L (2014) Effect of an on-body ergonomic aid on oxygen consumption during a repetitive lifting task. Int J Ind Ergon 44:39–44

Wind H, Gouttebarge V, Kuijer PPFM, Sluiter JK, Frings-Dresen MHW (2009) Complementary value of functional capacity evaluation for physicians in assessing the physical work ability of workers with musculoskeletal disorders. Int Arch Occup Environ Health 82:435–443

Chapter 12
QFD-Based Ergonomic Design of Drafting Table for Engineering Students: A Case Study

K.G. Durga Prasad, M.V. Prasad, Ch. Hima Gireesh and V.V.V.N.K. Chaitanya

Abstract Engineering design uses engineering drawing as the way to communicate and document ideas. Most of the universities in India have been offering engineering drawing as one of the courses for undergraduate engineering programs. The study of engineering drawing will enhance the imagination and descriptive ability of all engineering students and also help them analyze problems from different points of view. The accuracy of drawings and speed of execution are not only depending on the quality of the drawing instruments but also depending on human comfortability of drafting table and chair. This paper presents a methodology for conceptual design of ergonomic drafting table using Quality Function Deployment (QFD) with a view to satisfying the users (students) under ergonomic perspective. A case study is also presented in this paper to demonstrate the proposed methodology. Nordic Musculoskeletal Questionnaire (NMQ) survey has been conducted to understand the location of musculoskeletal discomforts. In order to identify and categorize the user needs and their potential influence on user satisfaction, Kano model analysis has been carried in the present study. On the basis of anthropometric data, NMQ survey and critical review of the results of Rapid Upper Limb Assessment (RULA), brain storming sessions have been conducted to explore the necessary design requirements. The principle tool of QFD is the House of Quality (HOQ), which maps the user needs and design requirements. While establishing the interrelationship matrix of HOQ, analytic network process (ANP) approach is adopted to address the issue of certain interdependencies between user needs and design requirements as well as the intra-dependencies among them. The proposed methodology paves the way in which the design team develops a drafting table with ergonomic design features to attain complete satisfaction of the users.

Keywords Ergonomics · Drafting table · Quality function deployment · Kano model analysis · ANP

K.G. Durga Prasad (✉) · M.V. Prasad · Ch. Hima Gireesh · V.V.V.N.K. Chaitanya
Department of Mechanical Engineering, G.V.P School of Engineering, Technical Campus, Gayatri Vidya Parishad College for Degree and P.G. Courses (A), Visakhapatnam, Andhra Pradesh, India
e-mail: dr.kgdp@gmail.com

P.K. Ray and J. Maiti (eds.), *Ergonomic Design of Products and Worksystems—21st Century Perspectives of Asia*, Managing the Asian Century,
https://doi.org/10.1007/978-981-10-5457-0_12

12.1 Introduction

In India, most of the universities offer engineering drawing as one of the courses for under graduate engineering programs. The students are expected to learn the course with a view to enhancing their knowledge on transforming data into graphic drawings which communicate the ideas to others. Though computer-aided design (CAD) approach plays a prominent role in engineering drafting work, the learning of manual drawing is necessary to understand the fundamentals and standardized practice are essential. Manual engineering drawing is generally viewed as providing a solid grounding in terms of knowledge and understanding of drafting as a concept. The conceptual understanding seemed to facilitate faster access to the learning required for CAD (McLaren 2008). Therefore, the practice of manual engineering drawing using drafting table and drawing instruments is necessary at least in the first year engineering program. The students of engineering may spend in front of drafting table for about 2 to 3 hours continuously. Though a numerous variety of drafting tables are available in terms of quality, durability, cost and aesthetic features, etc., most of the institutions prefer to provide drafting tables and chairs without serious look on ergonomic aspects than cost considerations. During the course of time, the students may subject to musculoskeletal disorders (MSDs) and thereby the learning outcome may not be achieved. In this context, a user-centered product development methodology has been formulated in the present work for designing ergonomic drafting table.

The consideration of ergonomic issues in product design has been increasing since the last several decades. Currently most of the manufacturers looking towards incorporating ergonomic aspects into their products intended to achieve complete satisfaction of the customers (users). It is essential to embed ergonomic needs of users in the conceptual phase of design process and then there is a scope of exploring necessary design attributes effectively. A methodological treatment is needed at this stage to translate user needs (UNs) into design requirements (DRs). The QFD is a customer (user) focused conceptual design approach which incorporates the UNs into the final product. In this paper, an attempt is made to propose a hybrid methodology using QFD for conceptual design of drafting table for engineering students. Kano model analysis is utilized in the proposed methodology for capturing and prioritizing the UNs. Prior to attempt for designing ergonomic drafting table, the intensity of human discomforts in the existing one is examined through NMQ survey and RULA. In order to address the issue of interdependencies between UNs and DRs as well as the intra-dependencies among them, analytic network process (ANP) is employed in the proposed methodology. The Kano-QFD-ANP hybrid methodology presented in this paper provides a concrete path for systematic mapping of UNs to DRs. The rest of the paper is organized as follows. The literature related to the techniques QFD, Kano and ANP is briefly reviewed in the Sect. 12.2. In Sect. 12.3, the proposed methodology is illustrated. In Sect. 12.4, a case study is presented to demonstrate the proposed methodology. The results and discussions are made in Sect. 12.5. Finally, the conclusions are discussed in Sect. 12.6.

12.2 Literature Review

12.2.1 Quality Function Deployment (QFD)

QFD is a customer focused quality engineering technique, which was developed in Japan in the late 1960s by Prof. Yoji Akao and Prof. Shigeru Mizuno with a view to incorporating customers' needs into the design of a product before it was manufactured. It is a proactive product development approach and it can be referred to as designed-in quality rather than traditional inspected-in quality (Chan and Wu 2002; Hashemi et al. 2015). The QFD process initiated with capturing the voice of customer and it can be used to measure customer satisfaction (Durga Prasad et al. 2008). The principal tool of QFD is the House of Quality (HOQ) which is a kind of conceptual map that provides the means for inter-functional planning and communications (Hauser and Clausing 1998). The HOQ appears like a normal house consisting of number of components namely, customer needs, design requirements, priority structure of customer needs, interrelationship between customer needs and design requirements, and the prioritized design requirements. The HOQ serves as means of accurate and exact translation of customer needs into design requirements.

QFD has been applied for different purposes to various fields such as automotive, manufacturing, construction, supply chain management, service, electronics, marketing, education, food industry, software, research and development, health care, ergonomics, etc. (Chan and Wu 2002; Sharma et al. 2008; Xu et al. 2010; Sivasamy et al. 2016). Currently, the use of QFD is increasing in the area of ergonomics. An ergonomic product may be regarded as a high-quality product because the quality as well as ergonomics aims at meeting complete satisfaction of the customers (Marvin et al. 2003). In the ergonomic product development process, it is important to match the human needs systematically with the product characteristics, which can be achieved by using QFD. The knowledge of ergonomics and the use of QFD are essential tools in designing high-quality products (Bergquist and Abeysekera 1996). Marsot (2005) emphasized that QFD is a methodological tool which is indeed capable of providing a solution to the problems of integrating ergonomics into product design. Guedez et al. (2001) employed QFD for ergonomic design of containers that will be used in flexible manufacturing systems. Evers et al. (2009) stated that QFD is a useful method in designing ergonomically more comfortable hand tools. Horvath et al. (2013) illustrated the adoptability of QFD for the ergonomic design of sitting furniture. Lin et al. (2016) presented a design methodology for developing ergonomic over bed table for bedridden patients using QFD.

While implementing QFD methodology, the key prerequisite is the understanding of customer needs and its potential influence on a product. To this end, Kano model analysis can help design team for analyzing customer needs and gaining profound understanding of customer satisfaction. The analyzed customer needs through Kano technique guides the design team through the conceptualization, creation, and realization process of a new product or a new version of an existing product. The Kano model analysis involved in QFD has a positive effect on product innovation (Shen et al. 2000). QFD and the Kano model can be integrated effectively to identify customer needs more specifically and to yield maximum customer satisfaction (Chaudha et al. 2011). Another significant aspect in QFD methodology is the establishment of interrelationship matrix of HOQ through systematic mapping of design requirements to customer needs. Many researchers have contributed to propose effective models of HOQ since 1960s, but most of the models assume the relationships between customer needs and design requirements are simply linear. Sometimes, customer needs and design requirements are not linear and independent. There may be certain interdependencies between customer needs and design requirements as well as the intra-dependencies among them. This issue is not considered in the traditional QFD. As the ANP technique handles the inner dependencies and outer dependencies and it may transform qualitative judgments into quantitative values (Lee et al. 2008), the involvement of ANP in QFD methodology may obviate the difficulty of establishing interrelationship matrix (Subbaiah et al. 2011). In the present work, Kano model analysis and ANP are integrated in the QFD process to capture voice of user and establish interrelationship matrix of HOQ, respectively. These two techniques are discussed briefly in the following paragraphs.

12.2.2 Kano model analysis

Kano model is a user preference classification technique, which was proposed by the Japanese professor Noriaki Kano and his colleagues in the 1980s. It is a useful tool for understanding customer needs and their impact on customer satisfaction (Wang and Ji 2010). According to Kano model, customer needs can be classified into following six categories such as attractive (A), must-be (M), one-dimensional (O), indifferent (I), reverse (R), and questionable (Q) (Chaudha et al. 2011; Gupta and Srivastava 2011). The execution of Kano model analysis consists of the following steps (Matzler and Hinterhuber 1998).

Step 1: Identification of the customer's needs:

The customer's needs are usually identified by conducting individual interviews with the customers or focus group discussions with the members who already know the product or service. The individual interviews are helpful in identifying one-dimensional needs, while focus group discussions help to identify attractive needs (Shiba et al. 1993).

Step 2: Developing and administering Kano questionnaire:

Kano questionnaire helps to categorize customer's needs into six Kano categories. The questionnaire examines each customer need with a pair of questions in functional and dysfunctional forms. There are five possible answers for each pair of questions: like, must-be, one-dimensional, neutral, live with, and dislike. On the basis of customer responses to both questions, the customer need is classified as one of six Kano categories for that customer by checking the Kano evaluation table (Berger et al. 1993).

Step 3: Evaluation and interpretation of the results:

According to the frequency of responses to the questionnaires, the expectation levels of customers' needs can be evaluated and interpreted. In order to estimate the average impact on customer satisfaction, Berger et al. (1993) proposed customer satisfaction index (CS). It is calculated by dividing the sum of frequencies of attractive f (A) and one-dimensional needs f (O) with the sum of the frequencies of attractive f (A), one-dimensional f (O), must-be f (M) and indifferent needs f (I). The mathematical expression for CS is given below.

$$CS_i = \frac{f(A) + f(O)}{f(A) + f(O) + f(M) + f(I)}$$

The values of CS lie between 0 and 1. The values close to 1 indicate greater satisfaction while the values close to 0 indicate lower satisfaction. The strategic implications of Kano model analysis are: fulfill all must-be needs, be competitive regarding one-dimensional needs, standout with regard to attractive needs, do not spend time or money for developing requirement that is perceived as indifferent, and avoid reverse requirements (Dominici et al. 2013).

Most of the researchers attempted to integrate Kano model in the QFD methodology with a view to developing products or services to meet the maximum satisfaction of the customers. Tontini (2007) believed that the integration of the Kano model in QFD can allow innovative requirements to receive the necessary attention in the product development process. Gupta and Srivastava (2011) dealt systematically how Kano's model of customer satisfaction can be integrated into QFD. Hashim and Dawal (2012) presented the joining of Kano model and QFD to improve the school workshop's work station design for adolescent in terms of ergonomic and users need. Taifa and Desai (2015) tried to show on the way Kano model can be integrated with QFD and ergonomics principles as the best way to capture user's requirements for achieving high satisfaction through an illustration of ergonomic design of class room furniture. Therefore, all the attempts made by the researchers on the issue of integrating Kano model and QFD reveal that the integrated approach is essential to attain total customer satisfaction in respect of products and services.

12.2.3 Analytic network process

Thomas Saaty developed analytic hierarchy process (AHP) as a decision-making method in the 1970s. The AHP technique constructs a decision-making problem in various hierarchies as goal, criteria, sub-criteria, and decision alternatives. It provides decision-makers with a way to transform subjective judgments into objective measures. In addition to AHP, ANP (analytic network process) also developed by Saaty is a generic form of AHP that allows for more complex, interdependent relationships, and feedback among elements in the hierarchy. ANP has been used in several decision-making applications in the last decade, especially in the study of risk and uncertainty (Sipahi and Timor 2010). The ANP is a coupling of two phases. The first phase consists of a control hierarchy or network of criteria and sub-criteria that control the interactions. The second phase is a network of influences among the elements and clusters. The network varies from criterion to criterion and a different super matrix of limiting influence is computed for each control criterion. Finally, each of these super matrices is weighted by the priority of its control criterion and the results are synthesized through addition for all the control criteria (Saaty 1999).

Several applications of ANP reported in the literature in various fields that the ANP is a better method to provide a flexible model for solving real-life situations. Specifically ANP is applied in QFD to address interdependencies and intra-dependencies. Pal et al. (2007) proposed a methodology by integrating ANP in QFD to determine and prioritize the engineering requirements of a cast part and to evaluate an appropriate rapid prototyping based route for tooling fabrication. Lin et al. (2010) adopted QFD and ANP to analyze the environmental production requirements in linguistic preferences. Subbaiah et al. (2011) developed customer-driven product planning methodology using conjoint analysis and QFD-ANP. Soota et al. (2011) conducted a case study for the development of cars using synergistic combination of ANP with QFD to evaluate most satisfying design for customers.

12.3 Methodology

The hybrid methodology proposed in this work provides a user-centered design approach in which human factors are of central concern within the design process. The schematic representation of the proposed hybrid methodology is shown in

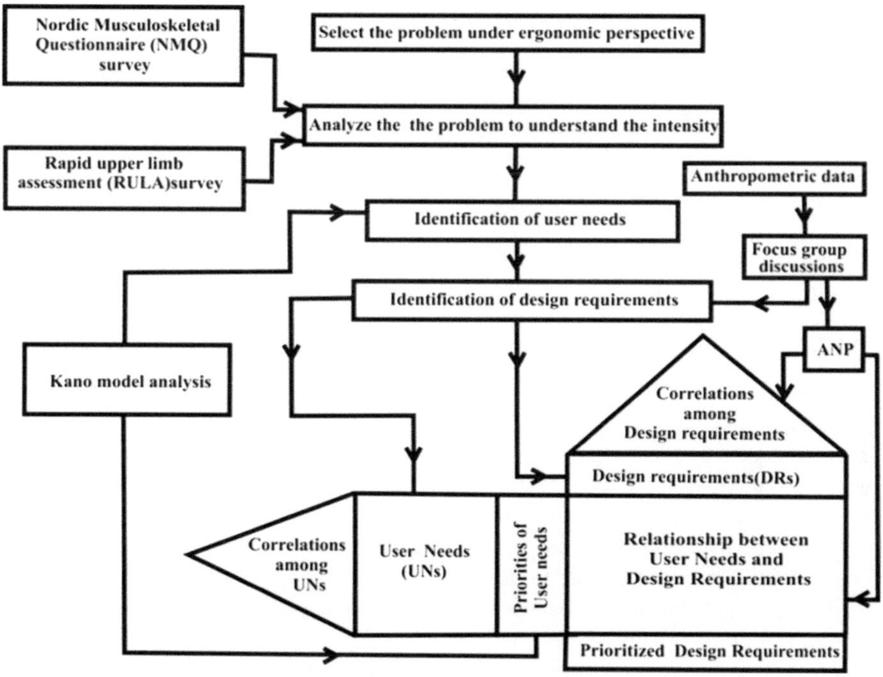

Fig. 12.1 Kano-QFD-ANP hybrid methodology for designing ergonomic product

Fig. 12.1. The step-by-step procedure to carry out the proposed methodology is discussed below:

Step 1: Select the problem under ergonomic perspective

Before attempting to study an ergonomic problem about a product, it is necessary to interact with the user group and conduct interviews. The user feedback survey reports are also helpful to identify the problems under ergonomics perspective.

Step 2: Analyze the problem to understand its level of intensity

It is necessary to analyze the problem of human discomfort in a product and then measure the level of intensity. The main three risk factors contributing to MSDs are force, repetition, and awkward postures. Any one or combination of these may contribute to the development of MSDs. The NMQ survey and RULA are the most commonly used tools of ergonomics research for identifying and analyzing the causes for MSDs. The NMQ is a tool for measuring MSDs and it can be used to assess a general level of discomfort across the workforce, regardless of the source of pain (Cornelius et al. 2001). The Rapid upper limb assessment (RULA) is a survey method used to understand the work-related upper limb disorders are reported. This method uses diagrams of body postures and three scoring tables to provide evaluation of exposure to risk factors.

Step 3: Identification of user needs

After discovering the level of discomforts in using a product, the exact user requirements have to be captured to proceed for new product development process. The Kano model analysis serves the purpose of capturing the voice of users (VOU) and helps to categorize the user needs.

Step 4: Prioritization of user needs

The priority structure of user needs to give scope on understanding about the intensity of desires about the product which is required for the design team while establishing design requirements. The Kano technique provides the user (customer) satisfaction index for all the user needs using Kano technique. The normalized values of satisfaction indices give the weightages for the user needs.

Step 5: Identification of design requirements

The technical portion of HOQ starts with the establishment of design requirements and completed with the determination of priority structure of the design requirements. As the users may not have sufficient knowledge of the individual functions of the product, they are unable to describe the features clearly in terms of technical specifications that are to be incorporated in the final product. But the core philosophy of QFD is the exact translation of user needs into design requirements. Therefore, each customer need may be translated into one or more design requirements by conducting brainstorming sessions among the focus group members.

Step 6: Determination of priority ratings of the design requirements

The following QFD-ANP procedure (Subbaiah et al. 2011) is employed to obtain the priority ratings of the design requirements.

(i) Establish the matrix (W_1) which shows the degree of relative importance of the DRs with respect to each UN.
(ii) Establish inner dependence matrix (W_2) of the UNs with respect to each UN
(iii) Establish inner dependence matrix (W_3) of the DRs with respect to each DR
(iv) Establish interdependent priority matrix (W_C) of the customer needs using the expression, $W_C = W_2 \times w$

where w = the matrix of priority ratings of customer needs

(v) Establish the interdependent priority matrix (W_A) of the design requirements using the expression, $W_A = W_3 \times W_1$
(vi) Determine the overall priorities of the design requirements (W^{ANP}) using the following expression $W^{ANP} = W_A \times W_C$.

The implementation of the methodology is demonstrated through a case study in the next section.

12.4 Case Study

A case study has been under taken in an engineering educational institution located in Visakhapatnam, Andhra Pradesh, India with a view to designing ergonomic drafting table for engineering students. The institution has been offering under graduation programs of four branches of engineering. The institution was started in the year 2011 and is running under the affiliation of a university with annual intake of 60 students per each engineering program. As per the curriculum of the university, first year students of all engineering branches have to learn engineering drawing as a compulsory course. The second and third year students of mechanical engineering need to learn machine drawing and production drawing courses. The students need to spend 3 hours continuously per week in front of drafting table during their practice sessions. Currently, the institution is arranging drafting tables provided with metallic frame and standard drawing boards and stools. The existing designs of drafting table and stool have been unaltered for years. During the study, it is noticed that students normally preferred to stand and draw and they have experienced discomforts due to awkward standing and sitting postures. Therefore, this issue has to be considered by the ergonomists to redesign the drafting table for the students and also the institutions should select drafting tables under prime consideration of ergonomic features as they have social responsibility toward student community.

The present study starts with conducting NMQ survey over 240 students of engineering who are the users of drafting table to understand the location of musculoskeletal discomforts. The survey revealed that the users have been experiencing severe pains in neck, upper back, middle back, lower back, buttocks and

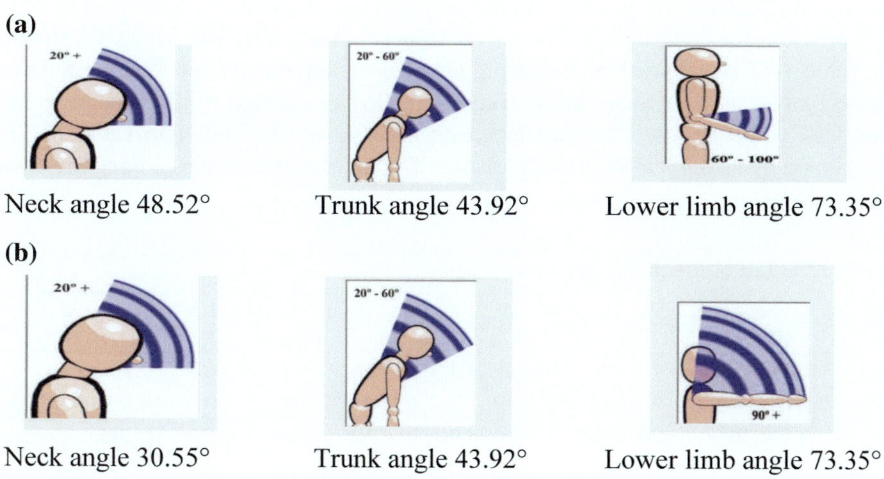

Fig. 12.2 **a** RULA for different postures of user at minimum awkward position. **b** RULA for different postures of user at maximum awkward position

mild pains in the other anatomical parts. After the NMQ survey, the photographs were taken at different postures of students during their drafting work for conducting RULA. Figure 12.2a, b shows the angles of neck, trunk, and lower limb at minimum and maximum awkward postures of user and the corresponding angles considered in RULA, respectively.

The RULA scores obtained are 5 and 6 for the postures in minimum and maximum awkward positions, respectively. The level of action is 3 and it indicates that the investigation and changes are required soon. Therefore, it is decided to proceed for product development. The total students of 127 consisting of 98 male and 29 female students were selected randomly, between 18–22 years of age, studying in different semesters and branches of engineering programs. The anthropometric data for these students have collected under defined standard positions (Chakrabarti 1997). In order to identify the UNs and their categorization, Kano questionnaire has been developed and administered to 50 students (35 male and 15 female) who actively involved in the survey. The user needs such as provision for placing drawing instruments (PPDI), convenient height of the table (CHT), flexible movement of drawing board (FMBD), comfortable chair (CC), movement of drafter to all corners of the table (MDCT) and low price table (LPT) are identified and are categorized as per the methodology discussed in Sect. 12.3. The Table 12.1 shows the list of UNs and Kano category, the user satisfaction index (USI) and weightages for all the UNs.

The outcome of brainstorming sessions held with the technical experts in the field of designing ergonomic products revealed seven design requirements for drafting table namely adjustable: height (AH), hinge mechanism (HM), locking system (LS), stability (ST), side shelf arrangement (SSA), chair with semi-back rest (CSBR), and separate provision for fixing drafter (SPFD). The matrix of the degree of relative importance of the DRs with respect to each UN (W_1) is developed by assuming that there is no dependence among the DRs and they are compared with respect to each UN provides the column Eigen vectors in respect of each UN. The matrix W_1 shown in Table 12.2 is obtained by preparing pair-wise comparisons between the DRs in respect of each UN. Saaty scale (Bayazit 2006) is adopted while preparing the pair-wise comparisons. The inner dependence matrix of UNs shown in Table 12.3 is developed by using pair-wise comparisons. The inner dependencies among the UNs

Table 12.1 List of user needs and Kano category

SI. No.	User needs (UNs)	A	O	M	I	R	Q	Total	Grade	USI	Weightage (w)
1	PPDI	15	20	10	2	3	0	50	O	0.7	0.175
2	CHT	20	15	5	5	5	0	50	O	0.7	0.175
3	FMDB	30	10	5	5	0	0	50	O	0.8	0.200
4	CC	0	40	10	0	0	0	50	O	0.8	0.200
5	MDCT	10	10	15	15	0	0	50	M, I	0.4	0.100
6	LPT	20	10	10	10	0	0	50	A	0.6	0.150

Table 12.2 Degree of relative importance of the DRs with respect to UNs (W_1)

DRs	UNs					
	PPDI	CHT	FMDB	CC	MDCT	LPT
AH	0.2288	0.2170	0.1036	0.1731	0.1115	0.1593
HM	0.1255	0.1016	0.1089	0.1229	0.1567	0.1880
LS	0.0710	0.1410	0.1905	0.0971	0.1337	0.1473
ST	0.1041	0.0839	0.1321	0.0973	0.0853	0.0878
SSA	0.2026	0.1913	0.1159	0.1536	0.0834	0.1494
CSBR	0.1210	0.0672	0.1810	0.2528	0.2040	0.1824
SPFD	0.1470	0.1978	0.1680	0.1032	0.2254	0.0858

Table 12.3 Inner dependence matrix of the user needs (W_2)

UNs	UNs					
	PPDI	CHT	FMDB	CC	MDCT	LPT
PPDI	0.4037	0.3899	0.2352	0.2714	0.2865	0.3907
CHT	0.1563	0.1580	0.1697	0.2342	0.0841	0.1718
FMDB	0.2375	0.2509	0.3447	0.2572	0.1327	0.2345
CC	0.1427	0.1447	0.1956	0.1672	0.0861	0.1389
MDCT	0.0309	0.0487	0.0328	0.0713	0.3835	0.0313
LPT	0.0774	0.0517	0.0704	0.0575	0.1415	0.0763

Table 12.4 Inner dependence matrix of the design requirements (W_3)

DRs	DRs						
	AH	HM	LS	ST	SSA	CSBR	SPFD
AH	0.4655	0.1722	0.1612	0.1140	0.1188	0.1199	0.2394
HM	0.2025	0.2532	0.2498	0.2186	0.1972	0.1995	0.1588
LS	0.1358	0.1976	0.2337	0.2397	0.1855	0.1986	0.1563
ST	0.0766	0.0887	0.1124	0.1700	0.1348	0.1337	0.1844
SSA	0.0292	0.0485	0.0766	0.0911	0.1595	0.1427	0.0882
CSBR	0.0372	0.0841	0.0748	0.0736	0.0859	0.0871	0.0908
SPFD	0.1960	0.2986	0.2344	0.2358	0.2612	0.2613	0.2250

are calculated by analyzing the impact of each UN on other UN. Similarly, the inner dependence matrix of the DRs is established by determining the inner dependencies among the DRs. The inner dependencies are obtained through analyzing the impact of each DR on other DR by establishing pair-wise comparisons. The Table 12.4 shows the inner dependence matrix of the DRs.

Then the interdependent priority matrix of the user needs (W_C) and interdependent priority matrix of the design requirements (W_A) are prepared and then the overall priorities of the design requirements are computed by using the formulae discussed in the Sect. 12.3. Table 12.5 shows the priorities of the design requirements.

Table 12.5 Priority ratings of the design requirements

SI. No.	Design requirements (DRs)	Priority ratings
1	Adjustable height (AH)	0.22
2	Hinge mechanism (HM)	0.22
3	Locking system (LS)	0.20
4	Stability (ST)	0.13
5	Side shelf arrangement (SSA)	0.10
6	Chair with semi-back rest (CSBR)	0.08
7	Separate provision for fixing drafter (SPFD)	0.26

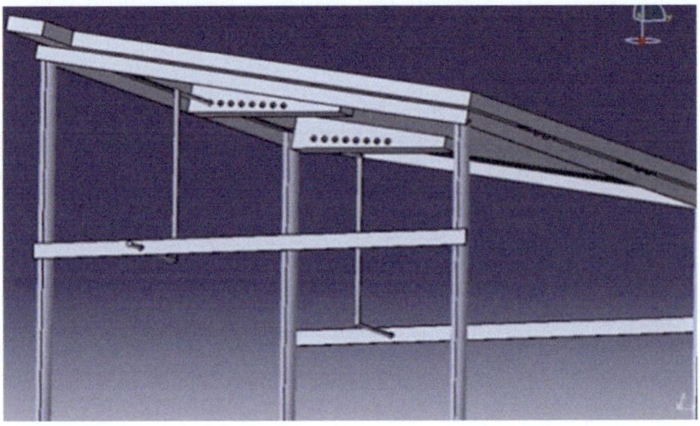

Fig. 12.3 Assembled view of ergonomic drafting table

12.5 Results and Discussion

The Kano model analysis along with ANP integrated in QFD is employed in this research to obtain the priorities of the design requirements to develop ergonomic drafting table for engineering students. From the Table 12.5, it is observed that the design requirement SPFD has highest priority value. Next the design requirements AH and HM have equal and higher priority values compared to other design requirements. The LS has the next priority rating. The rest of the DRs possess least priority values. Therefore, the design team has to focus primarily on the design requirements SPFD, AH, HM, and LS in an order. This priority order will play key role in the establishment of the parts to make ergonomic drafting table. The assembled view of the proposed drafting table is shown in Fig. 12.3. The RULA is conducted for the newly designed ergonomic drafting table and the score is decreased from 6 to 4. However, the score indicates that the risk with this table is low, small changes may be needed which can be resolved in further phases of design process. In the newly designed drafting table, a separate provision provided at left side corner of the table to fix the drafter so that the user can move the drafter to all corners of the table without any stress.

In order to provide adjustable height of the table, pipe in pipe with pin locking arrangement is provided in the new drafting table intended to make the adjustability to any height as desired by the users. Hinge mechanism and wedge with locking system are employed to satisfy the users in respect of providing flexible movement of drawing board. It is easy to fix the board at any convenient position with reference to user comfort.

12.6 Conclusions

The Kano, QFD-ANP hybrid methodology is developed in the present work for designing ergonomic drafting table for engineering students. The MSDs of users are decreased with the use of this ergonomic drafting table and thus the learning outcome of drawing course may be attained shortly as the discomforts and health issues are significantly reduced. The methodology provides a path for developing products which possess usability with quality. It reduces the gap between user's conception and designer's conception thereby allowing much scope to reflect the user needs in the final product. The methodology is highly suitable for participatory ergonomic culture and environment as its principal focus is user.

References

Bayazit O (2006) Use of analytic network process in vendor selection decisions. Benchmarking 13 (5):566–579

Berger C, Blauth R, Boger D, Bolster C, Burchill G, DuMouchel W, Pouliot F, Richter R, Rubinoff A, Shen D, Timko M, Walden D (1993) Kano's method for understanding customer-defined quality. Cent Qual Manage J 2(4):3–35

Bergquist K, Abeysekera J (1996) Quality function deployment (QFD) - A means for developing usable products. Int J Ind Ergon 18(4):269–275

Chakrabarti D (1997) Indian anthropometric dimensions. National Institute of Design Publishers, Paldi, Ahmedabad, India

Chan LK, Wu ML (2002) Quality function deployment: a comprehensive review of its concepts and methods. Qual Eng 15(1):23–35

Chaudha A, Jain R, Singh AR, Mishra PK (2011) Integration of Kano's model into quality function deployment (QFD). Int J Adv Manuf Technol 53(5):689–698

Cornelius KM, Turin FC, Wiehagen WJ, Gallagher S (2001) An approach to identify jobs for ergonomic analysis. In: Proceedings of the IIE 10th annual research conference, Dallas, Texas, 20–22 May 2001, pp 1–5

Dominici G, Palumbo F (2013) The drivers of customer satisfaction in the hospitality industry: applying the Kano model to Sicilian hotels'. Int J Leisure and Tourism Mark 3(3):215–236

Durga Prasad KG, Narayana Rao K, Padmavathi G, Venkata Subbaiah K (2008) Development of total quality engineering education model using QFD. J Eng Educ 21(4):1–6

Evers LFMK, Morel KPN, Eikelenberg NLW, Vink P (2009) Application of the QFD as a design approach to ensure comfort in using hand tools: can the design team complete the house of quality appropriately? Appl Ergon 40(3):519–526

Guedez V, Mondelo P, Hernandez A, Mosquera L (2001) Ergonomic design of small containers using the Quality Function Deployment (QFD). In: Proceedings of International conference on computer-aided ergonomics and safety. Maui, Hawaii, USA, 29 July–1 August 2001

Gupta P, Srivastava RK (2011) Customer satisfaction for designing attractive qualities of health care service in India using Kano model and Quality function deployment. MIT Int J Mech Eng 1(2):101–107

Hashemi N, Marzban M, Delavari S (2015) Quality function deployment: application to chemotherapy unit services. Middle East J Cancer 6(4):219–228

Hashim AM, Dawal SZM (2012) Kano model and QFD integration approach for ergonomic design improvement. Procedia-Soc and Behav Sc 57:22–32

Hauser JR, Clausing D (1988) The house of quality. The Harvard Bus Rev 3:63–73

Horvath PG, Kovacs Z, Denes L (2013) Ergonomic design using quality function deployment and design of experiments. For Prod J 63(7–8):257–262

Lee YT, Wu WW, Tzeng GH (2008) An effective decision-making method using a combined QFD-ANP approach. WSEAS Trans Bus Econ 5(12):541–551

Lin Y, Cheng HP, Tseng ML, Tsai JCC (2010) Using QFD and ANP to analyze the environmental production requirements in linguistic preferences. Expert Syst Appl 37:2186–2196

Lin T, Ekanayake A, Gaweshan LS, Hasan ZA (2016) Ergonomic product development of over bed table for bedridden patients. Comput-Aided Des Appl 13(4):1–11

Marsot J (2005) QFD: a methodological tool for integration of ergonomics at the design stage. Appl Ergon 36(2):185–192

Matzler Kurt, Hinterhuber Hans H (1998) How to make product development projects more successful by integrating Kano's model of customer satisfaction into quality function deployment. Technovation 18(1):25–38

Marvin EG, Gioconda Q, Bahill AT (2003) Improving product design using quality function deployment: the school furniture case in developing countries'. Qual Eng 16(1):47–58

McLaren SV (2008) Exploring perceptions and attitudes towards teaching and learning manual technical drawing in a digital age. Int J Technol Des Educ 18(2):167–188

Pal DK, Ravi B, Bhargava LS (2007) Rapid tooling route selection for metal casting using QFD-ANP methodology. Int J Comput Integr Manuf 20(4):338–354

Saaty TL (1999) Fundamentals of the analytic network process. In: Proceedings of the fifth international symposium of the analytic hierarchy process, Kobe, Japan, 12–14 August

Sharma JR, Rawani AM, Barahate M (2008) Quality function deployment: a comprehensive literature review. Int J Data Anal Tech Strat 1(1):78–102

Shen XX, Tan KC, Xie M (2000) An integrated approach to innovative product development using Kano's model and QFD. Euro J Innovation Manage 3(2):91–99

Shiba S, Graham A, Walden D (1993) A new American TQM, four practical revolutions in management. Productivity Press, Portland

Sipahi S, Timor M (2010) the analytic hierarchy process and analytic network process: an overview of applications. Manage Decis 48(5):775–808

Siva samy K, Arumugam C, Devadasan SR, Murugesh R, Thilak VMM (2016) Advanced models of quality function deployment: a literature review. Qual Quan: Int J Methodol 50(3): 1399–1414

Soota T, Singh H, Mishra RC (2011) Fostering product development using combination of QFD and ANP: a case study. J Ind Eng Int 7(14):29–40

Subbaiah KV, Durga Prasad KG, Narayana Rao K (2011) Customer-driven product planning using conjoint analysis and QFD-ANP methodology. Int J Prod Qual Manage 7(3):374–394

Taifa IW, Desai DA (2015) Quality function deployment integration with Kano model for ergonomic product improvement (Classroom furniture)—a review. J Multi Eng Serv Technol 2(9):2484–2491

Tontini G (2007) Integrating Kano model and QFD for designing new products. Total Qual Manage 18(6):599–612

Wang T, Ji P (2010) Understanding customer needs through quantitative analysis of Kano's model. Int J Qual Reliab Manage 27(2):173–184

Xu J, Xu X, Xie SQ (2010) A comprehensive review on recent developments in quality function deployment. Int J Prod Qual Manage, 6(4):457–494

Chapter 13
Ergonomics for Hand Screen Printing Workers: Cognitive Perception

S Shankar, J Karthick and R Naveenkumar

Abstract Hand screen printing, one among the prominent businesses in the textile sector of India lacks an ergonomic workplace and also stands behind in terms of research and development for improving the workers' workability. Workability of a person also depends on his/her attitude toward a problem, thus a cognitive study was done among 385 workers with some key factors like stress, job satisfaction, willingness to work. Study was done to find association of the above said factors with their age, work experience, and their physique (especially weight). Descriptive analysis and repeated measures were done using SPSS v20. Results of descriptive analysis showed that, 75% of the population, who reported non-willingness fell under 1–5 years' experience group. People with more than 10 years' experience reported high stress (41.7%) and job satisfaction (33.3%) which was slightly less than 1–5 years' experience group. Above results were backed up by the results of repeated measures test, as the mean of reported stress was found to be 'high' (CI 2.243–3.201) in the experience group of >10 years. Similar results were acquired for association with age. Major population who reported more stress and less satisfaction belonged to >40 years age group. But people who reported non-willingness to work were mostly of age 30–40 years. The mean of reported stress was almost similar for all the three (<30, 30–40, and >40 years) age groups, which gave an indication that the work environment needs an immediate alteration considering people's stress relief. Simple modifications might be incorporated to the workplace, which might ease pressure out of workers' head and eventually increase the workability.

Keywords Cognitive · Textile printing · Repeated measures · Job stress

S. Shankar (✉) · J. Karthick
Department of Mechatronics Engineering, Kongu Engineering College, Erode, TamilNadu, India
e-mail: shankariitm@gmail.com

R. Naveenkumar
Department of Mechanical Engineering, Kongu Engineering College, Erode, TamilNadu, India

P.K. Ray and J. Maiti (eds.), *Ergonomic Design of Products and Worksystems—21st Century Perspectives of Asia*, Managing the Asian Century,
https://doi.org/10.1007/978-981-10-5457-0_13

13.1 Introduction

The challenges of present work environment mostly affect workers' mind rather than causing physical discomfort. Main challenges on that account include, stress, mental collapse caused by overwork, and non-willingness to work or the intention to leave work (Boyas et al. 2012). Evidences were also demonstrated by many researchers showing that psychological aspects also impact employees' well-being (Mauno et al. 2013). People involved in more physical effort demanding jobs are also prone to higher job stress and poor job satisfaction. Work environment was said to account significantly for the mental stress of workers through factors like monotonous work, irregular timed work, and shift works (Panchal et al. 2015).

A number of factors like turnover, family, work–family conflict and job insecurity were assessed by many researchers, of which work–family conflict had huge impact on job stressors. Most of the works asserted that older workers handle stress better than the younger people (Boyas et al. 2012; Mauno et al. 2013), as they have realistic expectations which would ease pressure out of their heads and help them handle stress in a better way. Stress is found to adversely affect the quality of work from the workers (Golubic et al. 2009), and some positive association was also found with the above said effects and education.

Gender-related stress inducers were studied by some researches, and it was concluded that male's concern lies on the work-related factors and female had family as the primary concern for their well-being (Lindfors et al. 2007). Work and home stress was found to be associated with depression and anxiety (Fan et al. 2015; Wiegner et al. 2015). Though several factors were assessed by researchers, the major factor which had significant effect on stress was age, since it has both positive and negative impacts. Mitigations to get rid of or to minimize stress at work are minimal and mostly adopted by employers of greater business value. Small- and medium-level businesses completely lag behind in terms of research and development on these aspects. In spite of that, some researchers provided suggestions to managers of businesses to develop strategies to address and improve the quality of working conditions (Golubic et al. 2009).

Hand screen printing (Fig. 13.1), one of the prominent businesses among the textile sector in India, lacks proper research and development when ergonomics and workers' safety is concerned. Also awareness regarding the proper work environment is also less among the workers which also led to the employer's ignorance. So most of the people leave this work soon and switch over some other local jobs. The attributes demonstrated by most of the above-said researches go well with hand screen printing business. As people of age ranging from 19–55 years take this work for their income, their attitude toward work should be analyzed to improve the productivity in terms of quality and most importantly their safety and eventually their well-being.

This paper presents a perception on the behavioral aspects of the workers in hand screen printing job and its association with their workability. Main objective includes the assessment of mental distress and thereby discussing about the

Fig. 13.1 Screen printing job and its environment

declining trend of interest to take hand screen printing. On that account, a questionnaire survey was done following which the percentage of number of workers' report on job satisfaction and stress were evaluated to arrive at conclusions. Then the results were discussed considering previous findings and facts with main concern on the present scenario. As the case considered is of economic significance, discussion would revolve on development of the business by reducing the shortcomings.

13.2 Methodology

13.2.1 Workplace and Subjects

The workers print 400–900 m per day depending on the order received. The table length lies between 50–100 m and width of about 3 m. Mostly, the work in the industry was shared by 10–15 workers. Work hours on hand screen printing industry are ≥ 10 h. Three rest times were provided to the workers, two tea breaks and one lunch break. The printing is done with the help of screen which consists of rectangular steel frame and the steel frame is fitted with a stencil. The stencil has the desired pattern. The stencil frame is 1 m in length and 3 m in width, weighs around 5–10 kg. About 385 hand screen printing workers were considered for the study with different age and experience. The workers work for at least 10 h a day.

13.2.2 Protocol

Questionnaire survey was done among 385 (211 male and 174 female) workers from 41 hand screen printing industries located at Erode district in Tamil Nadu, India. Questionnaire included demographic details like age, experience, height, weight, etc.; details of worker's attitude toward the job, its nature and also had some details about their discomfort. As the scope of the present work was confined with cognitive aspects of worker's discomfort, descriptive tests and repeated measures were carried out for the association of factors like stress, job satisfaction and their willingness to work, with demographical details of the workers (age, experience). The above-said analyses were done using SPSS v20.

Discomfort level were stated in words and also given values for analyses purposes. The answer for the question about worker's job stress and satisfaction included three ratings like low, moderate, and high. The values for the same were 1, 2, and 3, respectively. Percentage of number of workers reporting discomfort was found and its association with the worker's age and experience were tabulated.

13.3 Results

Descriptive analysis (Crosstab) done to find the possible association of years of experience and age with that of factors stress, willingness, and job satisfaction was shown in Tables 13.1 and 13.3. From those results, it can be inferred that willingness to work in the HSP industry varies loads after 1 year of experience. Also willingness was almost similar for all age group of workers except for the age group of 30–40 years in which people reporting non-willingness were more (43.4%). Most of the population either had low job satisfaction or moderate satisfaction, and people who were highly satisfied were only few, among them majority belonged to 1–5 years' experience group.

Table 13.1 Association of Experience with psychological factors

Factors		Years of experience			
		1 year	1–5 years	5–10 years	>10 years
		%n	%n	%n	%n
Willingness	Yes	4.9	50.8	14.8	29.5
	No	25.0	75.0	0.0	0.0
Job satisfaction	Low	6.2	31.2	31.2	31.2
	Moderate	7.5	62.5	5	25
	High	0.0	44.4	22.2	33.3
Job Stress	Low	0	42.9	21.4	35.7
	Moderate	7.7	66.7	5.1	20.5
	High	8.3	16.7	33.3	41.7

Table 13.2 **a** Estimation of reported job satisfaction in association with experience. **b** Estimation of reported stress in association with experience

Experience (Years)	Mean	Std. Error	95% confidence Interval	
			Lower	Upper
(a)				
1	1.750	0.310	1.129	2.371
1–5	1.971	0.106	1.758	2.183
5–10	1.667	0.207	1.253	2.080
>10	1.889	0.146	1.596	2.181
(b)				
1	2.250	0.321	1.608	2.892
1–5	1.882	0.110	1.662	2.103
5–10	2.111	0.214	1.683	2.539
>10	2.000	0.151	1.697	2.303

Table 13.3 Association of Age with psychological factors

Factors		Age (years)		
		<30	30–40	>40
		%n	%n	%n
Willingness	Yes	30.4	33.4	36.2
	No	26.1	43.5	30.4
Job satisfaction	Low	30.8	34.6	34.6
	Moderate	29.2	34.4	36.5
	High	35.6	31.1	33.3
Job Stress	Low	40.4	36.8	22.8
	Moderate	29.1	34.0	36.9
	High	23.9	30.4	45.7

Among the people who reported high stress in their job, more numbers (45.7%) were found in the higher age group (>40 years). This case remained same for moderate stress, as 36.9% was the highest percentage among who reported moderate stress and they belonged to the >40 years age group. From the results for association of experience with that of job stress, it was found that 66.7% of people who reported moderate stress in their job belonged to the 1–5 years' experience group. 5–10 years' experience people who reported job satisfaction, stress were never the majority side, they were either minor or medium in their percentage distribution.

Repeated measures test (Estimation of marginal means) has shown (Tables 13.2a, b and 13.4a, b) that ≤ 1 year of experience people had higher mean (2.250) for reported job stress (CI 1.608–2.892) and 1–5 years' experience people had higher mean for reported job satisfaction (CI 1.758–2.183). The above results clearly indicate that ≤ 1 year experience people had more job stress and 1–5 years group had less job satisfaction. Similarly >40 years of age people reported more stress (CI 1.972–2.144) and <30 years of age people had less job satisfaction (CI 1.908–2.092) (Table 13.3).

Table 13.4 **a** Estimation of reported job satisfaction in association with age. **b** Estimation of reported stress in association with age

Age (Years)	Mean	Std. Error	95% confidence Interval	
			Lower	Upper
(a)				
<30	2.000	0.047	1.908	2.092
30–40	1.969	0.044	1.883	2.056
>40	1.978	0.043	1.894	2.063
(b)				
<30	1.897	0.048	1.803	1.990
30–40	1.947	0.045	1.858	2.035
>40	2.058	0.044	1.972	2.144

13.4 Discussion

Age remained a major factor for people's stress and their willingness to get involved in the job (Fan et al. 2015; Greenberg et al. 2015). Present work results supported the same, as the non-willingness to work was higher among >40 years old people than <30 years old (Table 13.3). Also the estimated mean acquired for both stress and job satisfaction shown that the former was higher and the latter was lower for the higher age group (>40). Also, the younger workers intend to leave the job than older workers, since their chance to get alternative jobs is far pleasant than the counterparts (Boyas et al. 2012). Further analysis by including emotion regulation and other strategies would provide better conclusions (Mauno et al. 2013), paving way for possible improvement in the work environment.

Burnout and exhaustion rates are common for people above 35 years of age and also for some women under the above-said age (Wiegner et al. 2015). In our study too, the rates went well with previous research findings showing that women had higher stress than men. Stress with respect to weight of individual worker had no significance similar to work of Travis et al. (Wakeham et al. 2015), where mental stress response was similar for both normal weight and obese people. As our case of study demands considerable physical tasks from each and every worker, there is a bit of contradiction over other jobs where people of higher age handle stress with their acquired coping skills than the younger workers (Boyas et al. 2012).

Education and awareness about people's mental distress and causes might reduce perceived work loss (Fahey et al. 2016). Many researchers concluded that a good working environment increases the productivity since it was found to enhance the workability of the individual workers (Panchal et al. 2015). Our work environment lags in design aspect and provokes serious concern regarding the safety equipment and tools used. This might induce thoughts of risks for the workers, which will eventually reduce their willingness to work in the environment. Individual behavior, like complete detachment from work during leisure time, might be taken with some practices and considerable reduction in stress could be achieved (Sonnentag and Fritz 2015).

Further detailed study about the depression status might provide information regarding the cardiovascular reactivity (Wang et al. 2016) among the workers, since severity was not assessed in detail in the present study. Assessment of individual workers' capacity to deal with stress (Ohta et al. 2015) might be done in future, so that improvement in the environment could be achieved in a better way, leading to the improvement in workability.

13.5 Conclusion

This investigative study, on the cognitive distress of textile printing workers, motivated the need for a change in the working environment. Also the distress level status with respect to age and experience urged the need for a regulative work procedure model with special consideration for age factors. The present study found that people with experience of 1–5 years reported moderate stress (66.7%) and moderate job satisfaction (62.5%), which was greater than most other percent of the work groups. A thoughtful act by the managers or employers to retain fresher and less experienced people would enhance the productivity in the business. Moreover, workability of the employee holds the key for any development in a business. Hence, environment and workers' safety, stressing on aspects like reduced work pressure, lesser risk, nil health hazards, etc., should be enhanced.

Acknowledgements The Authors would like to thank Science and Engineering Research Board (SERB), India, for their financial support and also the workers of hand screen printing industry for attending survey and experimental trails patiently.

References

Boyas J, Wind LH, Kang S-Y (2012) Exploring the relationship between employment-based social capital, job stress, burnout, and intent to leave among child protection workers: an age-based path analysis model. Child Youth Serv Rev 34:50–62

Fahey N, Soni A, Allison J, Vankar J, Prabhakaran A, Simas TM, Byatt N, Phatak A, O'Keefe E, Nimbalkar S (2016) Higher levels of education mitigate the relationship between perceived stress and common mental disorders among women in rural India: results of a cross-sectional study. Ann Glob Health 82:560

Fan L-B, Blumenthal J, Watkins L, Sherwood A (2015) Work and home stress: associations with anxiety and depression symptoms. Occup Med 65:110–116

Golubic R, Milosevic M, Knezevic B, Mustajbegovic J (2009) Work-related stress, education and work ability among hospital nurses. J Adv Nurs 65:2056–2066

Greenberg PE, Fournier A-A, Sisitsky T, Pike CT, Kessler RC (2015) The economic burden of adults with major depressive disorder in the United States (2005 and 2010). J Clin Psychiatry 76:155–162

Lindfors P, Meretoja O, Töyry S, Luukkonen R, Elovainio M, Leino T (2007) Job satisfaction, work ability and life satisfaction among Finnish anaesthesiologists. Acta Anaesthesiol Scand 51:815–822

Mauno S, Ruokolainen M, Kinnunen U (2013) Does aging make employees more resilient to job stress? age as a moderator in the job stressor–well-being relationship in three finnish occupational samples. Aging & Ment Health 17:411–422

Ohta M, Higuchi Y, Kumashiro M, Yamato H, Sugimura H (2015) Work improvement factors for the amelioration of work ability, with a focus on individual capacity to deal with stress in an IT company. J UOEH 37:23–32

Panchal S, Joshi HL, Kumar U (2015) Role of work environment in stress and mental health. J Res: THE BEDE ATHENAEUM 6:98–103

Sonnentag S, Fritz C (2015) Recovery from job stress: the stressor-detachment model as an integrative framework. J Organ Behav 36:S72–S103

Wakeham T, Harwood M, Carter J, Durocher J (2015) Obesity and neural cardiovascular responses to mental stress in humans. FASEB J 29: 806.805

Wang M-Y, Chiu C-H, Lee H-C, Su C-T, Tsai P-S (2016) Cardiovascular reactivity in patients with major depressive disorder with high-or low-level depressive symptoms a cross-sectional comparison of cardiovascular reactivity to laboratory-induced mental stress. Biol Res Nurs 18:221–229

Wiegner L, Hange D, Björkelund C, Ahlborg G (2015) Prevalence of perceived stress and associations to symptoms of exhaustion, depression and anxiety in a working age population seeking primary care-an observational study. BMC Fam Pract 16:1

Chapter 14
Effectiveness of Classroom Interactions in Engineering Colleges

L Princy, S Swapnesh, R Rajesh and K Hridya

Abstract The teaching–learning environment is being influenced by technological advances in information and communication technology domain worldwide. This needs to be examined from the traditional teaching–learning methods adopted in many of the institutions. This paper is aimed to study the role of various classroom factors across diverse samples on classroom interactions and to model a causal relationship between the factors and classroom outcomes. This study is based on a man–machine model for classroom. As per this model the classroom factors are classified into five distinct dimensions, i.e. 'General Environment', 'Technology', 'Work Environment', 'Pedagogy' and 'Time'. The effects of these factors are analysed on two outcome dimensions, i.e. 'Academic Performance' and 'Ergonomic Strain'. For this purpose, a questionnaire survey is conducted among the students of an engineering college in the state of Kerala, India. The questionnaire consists of three parts; Part-A of questionnaire uses a 5-point Likert scale, part-B uses check list type questions, while part-C consists of multiple choice questions. Factor analysis is done on the work system factors through SPSS Promax procedure with Kaiser Normalization. A structural equation modelling is used for building the causal relationship between classroom dimensions and outcome dimensions. Based on the results from the study, recommendations are made for classroom modification for enhanced interaction and enriching experience.

Keywords Classroom interactions · Workspace · Questionnaire survey · Factor analysis · Structural equation modelling

---

L. Princy (✉) · S. Swapnesh · R. Rajesh · K. Hridya
Department of Mechanical Engineering, Rajiv Gandhi Institute of Technology, Kottayam, Kerala, India
e-mail: princylukose91@gmail.com

P.K. Ray and J. Maiti (eds.), *Ergonomic Design of Products and Worksystems—21st Century Perspectives of Asia*, Managing the Asian Century,
https://doi.org/10.1007/978-981-10-5457-0_14

14.1 Introduction

Developments in information and technology can be found in every field of our life at any time. Ignoring these developments is almost impossible. Education is also trying to catch up with these developments. The teaching–learning environment is being influenced by technological advances in information and communication technology domain worldwide. The growing complexity of technological systems and the quest for development now makes it imperative for educational institutions to shift from the existing method of teaching to accommodate use of technologies.

The nature of educational practices is getting transformed through initiatives such as outcome-based accreditation, ensuring richer learning experience through open educational resources, providing mass and flexible learning through distance learning programmes and satellite-based course, evolving classroom configurations, and advances in interaction tools, devices and platforms (Karal and Balicekapili 2010). This needs to be examined from the traditional teaching–learning methods adopted in many of the institutions. For effective learning of the students new learning patterns, evolving education technologies and methodologies needs to be better understood by all stakeholders. Hence the role of significant factors that influences teaching–learning environment needs to be explored.

The objective of this paper is to study the role of various classroom factors across diverse samples on classroom interactions and to model the causal relationship between work system dimension and outcome measures. Preliminary findings from a questionnaire survey conducted in an engineering college in Kerala and the results and discussion from the SEM are presented in this paper.

14.2 Literature Review

Today's educational systems have diverse workplace characteristics (Raju and Gupta 2012), and hence it is important to collect feedback from students regarding the existing systems. It is necessary to resolve the issues pertaining to current classroom and educational practice (O'bannon and Thomas 2014) for improving the students' performance. Worksystem outcomes are numerous and needs to be carefully studied to understand the role and effect of the different worksystem dimensions on students' performance. A modified man–machine model (Rajesh and Reena 2015a) is useful for analysing the ergonomic/human factors for digital class room (Bridger 2008) (see Fig. 14.1). The components of the worksystem are work environment, general environment, pedagogy, technology and time through which the teaching–learning occurs (Rajesh and Reena 2015a; Rajesh et al. 2015; Beauchamp and Kennewell 2010; Mohammed 2013). This model gives an insight to man–machine interactions, which are made feasible by a set of interactions through display, sensory mechanism, central processor, effector mechanism and controls. A good integration between different classroom components influences the

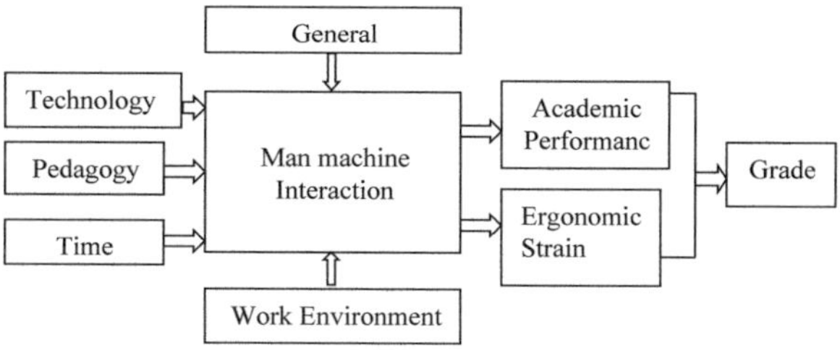

Fig. 14.1 A man–machine model (Modified: Bridger 2008; Rajesh and Reena 2015a)

nature of teacher–student interaction. The immediate positive outcomes include enhanced active learning, and lower review effort, while negative outcomes include poor clarity, boredom or disinterest, mental and physical strain (Rajesh et al. 2014; McGrath and Van Bergen 2015; Decker et al. 2007; Griggs et al. 2009). The long-term positive outcome is better grades, carrier opportunities and course expertise, while long-term negative outcome include psychological stress.

There is a need to examine the effects of technology and evolving concept of workspace within Indian educational campus. The technology-enabled class room calls for not only the integration of work system dimensions, i.e. course, technology, environment and human factors, but also needs to consider the issues pertaining to technology adoption (Murray and Zvoch 2011). Also the human factors such as age, experience, motivation and course factors such as course type, content and scheduling needs to be explored (Rajesh and Reena 2015b) so that the nature and impact on the interactions with teaching–learning environment can be better understood. Rajesh et al. (Mohammed 2013) and Rajesh and Reena (2015b) studied three classroom components, i.e. 'Technology', 'General environment' and 'Work environment', in an engineering college in Kerala, India. The focus is to examine the role of 'pedagogy' and 'time' in addition to the above three. The task set is to perform a questionnaire survey on the student's perception of classroom environment and analyse the factors and dimensions through factor analysis. In order to control a process it is necessary to have a model of that process (Conant and Ashby 1970). The structural equation modelling helps in establishing a causal relationship between the classroom dimensions and the outcome of teaching–learning process (Rajesh and Reena 2015b).

14.3 Methodology

The overall study methodology is shown in Fig. 14.2. A questionnaire has been developed, based on the literature review to collect responses so as to evaluate a man–machine model with contributing factors identified as 'General Environment', 'Technology', 'workspace', 'Pedagogy' and 'Time' and two outcome measures as 'Academic Performance' and 'Ergonomic Strain'. The questionnaire was distributed

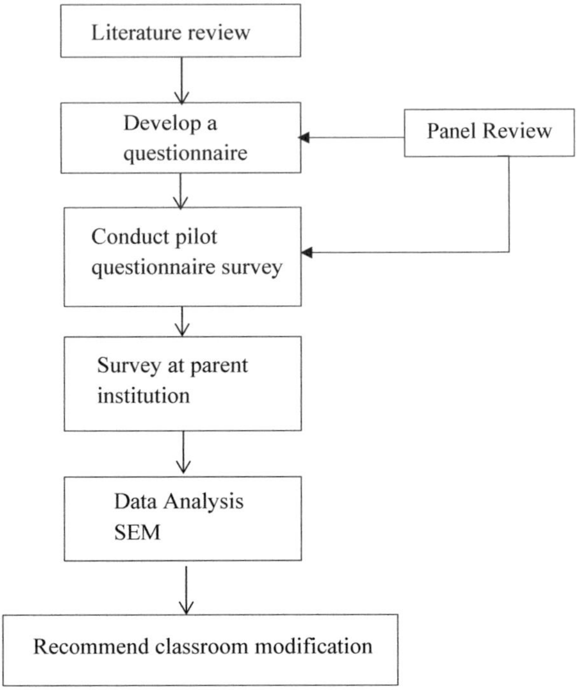

Fig. 14.2 Study methodology

to the students in one of the engineering colleges in Kerala; the institution chosen was purely based on convenience. The population size is ~1500 students from six engineering branches each having four batches corresponding to first, second, third and fourth year and with a class strength of 60–70 students. A random sampling was done with a confidence interval of 0.05 from the population and at 95% confidence level. Direct method is adopted and both online version and printed questionnaires were issued among students in various classes. A total of 347 samples were collected.

Data analysis for the survey includes questionnaire validity through Cronbach alpha, descriptive statistics of each of the questionnaire component, and Confirmatory factor analysis done using principal component analysis with promax rotation method in SPSS software (Version 20) on the worksystem factors and outcomes measures. The structural equation model is built based on the path diagrams generated for classroom factors and for outcome measures. AMOS software is used for building and testing the SEM. The base SEM is improved by considering p-values of the regression coefficients and use of modification index of the paths through an iterative procedure. The model fitness is evaluated using p-value, GFI, CFI and RMSEA.

14.3.1 Questionnaire Description

The questionnaire consists of three parts A, B and C. Part-A consists of 34 questions for the assessment of dimensions of current teaching–learning environment that accounts for factors such as connectivity and Internet, space, lighting, communication facility, etc., on a 5-point Likert scale. Part-B is a checklist-based questionnaire with 10 questions eliciting responses from on the typical issues faced in the current environment. Part-C consists of 14 questions that assess the outcome measures such as active learning achieved, academic grades and health issues.

14.4 Results and Discussion

14.4.1 Questionnaire Survey

Table 14.1 shows the mean responses of the Part-A containing the dimensions of the teaching–learning environment. Some of the factors had a mean rating value below 3 (i.e. below good rating) such as 'internet connectivity' (Q3—Availability of connectivity & internet facility for students in class room and Q4—Use of connectivity & internet facility by students within class-room), 'furniture design' (Q7—Availability of class room furniture and its features to accommodate student's requirements), 'storage facility' (Q19—Availability of storage facility for digital resources and tools and Q20—Availability of safety and security features for digital resources and tools).

The pie chart shows (Fig. 14.3) the students' feedback regarding the issues prevalent in the current teaching–learning environment. The Part-B complements the factors in Part-A. The common interactions tools mainly used include board and projectors and students rely on writing down lecture notes as a mode of recording notes (Fig. 14.3a, b). Even though both the faculty and students does have knowledge to use information and communication technology (Table 14.1, Q2, Q6) the use of ICT is limited one of the factors being technology-dependent operational issues (Fig. 14.3c).

The spatial and general environment is examined through questions 4–8 of Part B (see Fig. 14.3d–h). Some of the issues highlighted are congested space, long pathways, narrow classrooms, poor storage space, heat stress, etc. There appears to be low interactions leading to outcomes such as boredom and poor concentration (Fig. 14.3j). With respect to safety/security issues prevalent in classroom most of the respondents indicated 'misplacement' and 'material damage' as a cause for concern (Fig. 14.3i).

Table 14.1 Descriptive statistics of survey

Label	Dimensions	Mean	Std. Deviation	Label	Dimensions	Mean	Std. Deviation
Q1	Avail_ICT	2.71	1.111	Q18	personal belongings' storage	2.74	1.140
Q2	Fact_ICT	3.16	1.023	Q19	digital tools' storage	2.36	1.083
Q3	Use_fact_ICT	2.99	1.087	Q20	availability of safety	2.44	1.088
Q4	Avail_internet	2.43	1.211	Q21	match b/w ICT and course	2.86	0.961
Q5	Quality_internet	2.60	1.214	Q22	influence of course content	2.87	0.913
Q6	Know_stud_ICT	3.41	1.008	Q23	quality of class schedule	3.08	0.888
Q7	Use_stud_ICT	2.70	1.196	Q24	course duration	2.76	0.989
Q8	Stud.intract due to ICT	2.66	1.061	Q25	Student Int_exam	3.43	0.975
Q9	Readiness to use ICT	3.04	1.082	Q26	Student Ext_exam	3.82	1.039
Q10	Avail space	3.18	1.044	Q27	Faculty Int_exam	3.59	0.933
Q11	Layout	3.18	.991	Q28	Faculty Ext_exam	3.59	0.976
Q12	Flexi_classroom	3.10	1.011	Q29	Discussion encouragement	3.29	0.990
Q13	Avail_furniture	3.09	1.019	Q30	Discussion initiative	3.13	0.899
Q14	Flexi_stud_space	3.06	0.960	Q31	adequacy of allowance	2.82	1.061
Q15	Quality_light	3.32	1.012	Q32	refreshment quality	2.59	1.126
Q16	Quality sound	2.92	1.042	Q33	adequacy of duration	3.08	0.938
Q17	Phy. Envir. Quality	2.91	1.070	Q34	Class periods appropriate	3.10	0.844

14.4.2 Factor Analysis and Structural Equation Modelling

Confirmatory Factor analysis was carried out. The extraction of components has been done by principal component analysis method (PCA). The method of rotation chosen is promax with Kaiser Normalization, since correlations have been found between the variables. From the factor analysis, it has been found that the contributing factors from the initial man–machine model were regrouped into six, i.e.

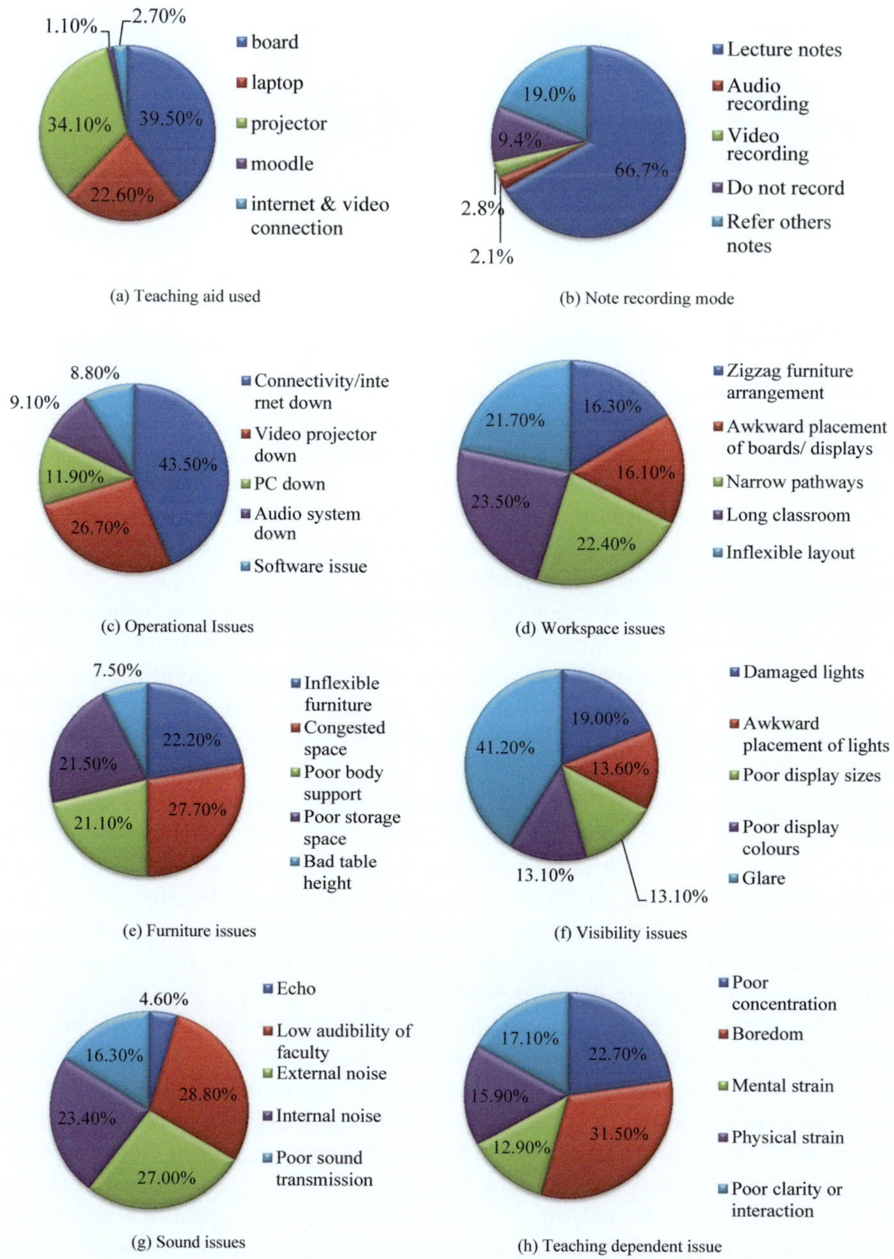

Fig. 14.3 **a** Teaching aid used, **b** Note recording mode, **c** Operational issues, **d** Workspace issues, **e** Furniture issues, **f** Visibility issues, **g** Sound issues, **h** Teaching dependent issue, **i** Environment issues, **j** Safety issue

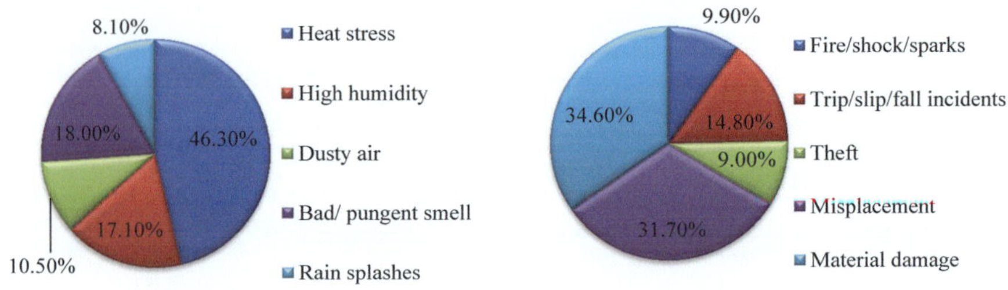

(i) Environment issues (j) Safety issue

Fig. 14.3 (continued)

contributing factors identified as ‘Technology’, ‘Workspace Environment’, ‘Pedagogy’ ‘Syllabus’ and ‘Time’ and outcome measures into four as ‘Academic Performance’, ‘Effort’, ‘Grades’ and ‘Ergonomic Strain’. The pattern matrix obtained from the factor analysis is shown in Fig. 14.4. The sampling adequacy has been determined using Kaiser–Meyer–Olkin MSA which is found to be 0.867 and Bartlett’s test of sphericity was significant, $\chi2(780) = 6530.626$, $p < 0.0001$. This means that the sample size is very adequate.

The structural equation modelling was done using AMOS software. In the measurement model, a confirmatory factor analysis was done to verify the contribution of independent variables to concerned factor. The measurement model was modified till necessary conditions of model fit were satisfied. Modified measurement model is given in Fig. 14.5 and model fit parameters of the measurement model is given in Table 14.2.

The structural model is used to determine the interactions within a classroom worksystem. The initial hybrid model developed for the selected sample is shown in Fig. 14.6 and the modification history of the model is given in Table 14.3. The final structural equation model for classroom worksystem developed from the responses given by the sample is shown in Fig. 14.7. The direct effect such as ‘Technology’—‘Performance’, ‘Effort’—‘Grade’, etc., were removed as the interactions were found insignificant ($p = 0.000$).

In this study, it has been found that among the six worksystem dimensions workspace environment and course content indicated by ‘workspace’ and ‘syllabus’, respectively, has greater effect on students’ performance than other classroom factors. Classroom workspace (0.181) is the main factor contributing to ergonomic strain. The ergonomic strain has a negative impact on class performance (−0.285) and grades (−0.042). And in turn, poor class performance leads to increased effort towards studying (−0.312). Self-study hours spent to comprehend the classroom teachings after classroom engagements are found to be very high (0.897). Layout (i.e. arrangement and location) of classroom facilities (i.e. tables, chairs, shelf, boards, projectors) for student–teacher interaction and learning (0.782) and flexibility towards seating arrangement or layouts within the current classroom space (0.770) are having highest loading on workspace environment component.

Pattern Matrix[a]

	Component									
	Technology	Workspace	Storage	Syllabus	Pedagogy	Time	Performance	Effort	Grades	Strain
avail_ict	.683									
fact_ict	.546									
use_fact_ict	.808									
avail_internet	.774									
qlty_internet	.681									
use_stud_ict	.820									
ict_interactn	.609									
avail_space		.722								
layout		.890								
class_flexi		.838								
avail_furniture		.768								
flexi_space		.724								
qlty_light		.629								
qlty_sound		.462								
storage_personal			.675							
digital_storage			.626							
digital_safety			.608							
ict_course				.773						
course_influnc				.740						
qlty_schedule				.656						
stud_int_exam					.660					
stud_ext_exam					.817					
fact_int_exam					.814					
fact_ext_exam					.843					
discn_encourag					.652					
allowance						.803				
refrshment						.827				
sub_understnd							.844			
solve_abiliity							.779			
cont_retain							.791			
stdy_effort								.750		
stdy_time								.785		
examprep_time								.748		
int_evaltn									.869	
ext_evaltn									.876	
strs_head										.757
strs_hand										.683
strs_posture										.716
strs_head_light										.742
irritation										.612

Fig. 14.4 Pattern matrix

Availability of connectivity and Internet facility for students in classroom (0.805) and quality of connectivity and internet facility (0.752) are having highest loading on technology component. It can be substantiated by the mean (SD) students response values of 2.43(1.21) and 2.6(1.21), respectively, from the results of data analysis of part-A of questionnaire (Table 14.1). An improvement in this dimension shall have a positive effect on interactions in classroom environment. The storage dimension is weak with respect to availability of storage facility and security for digital resources and tools [mean value (SD) for which is 2.36(1.08) and 2.44(1.09) respectively (Table 14.1)]. As a result of this cases of misplacement/theft and material damage have been reported (Fig. 14.3i).

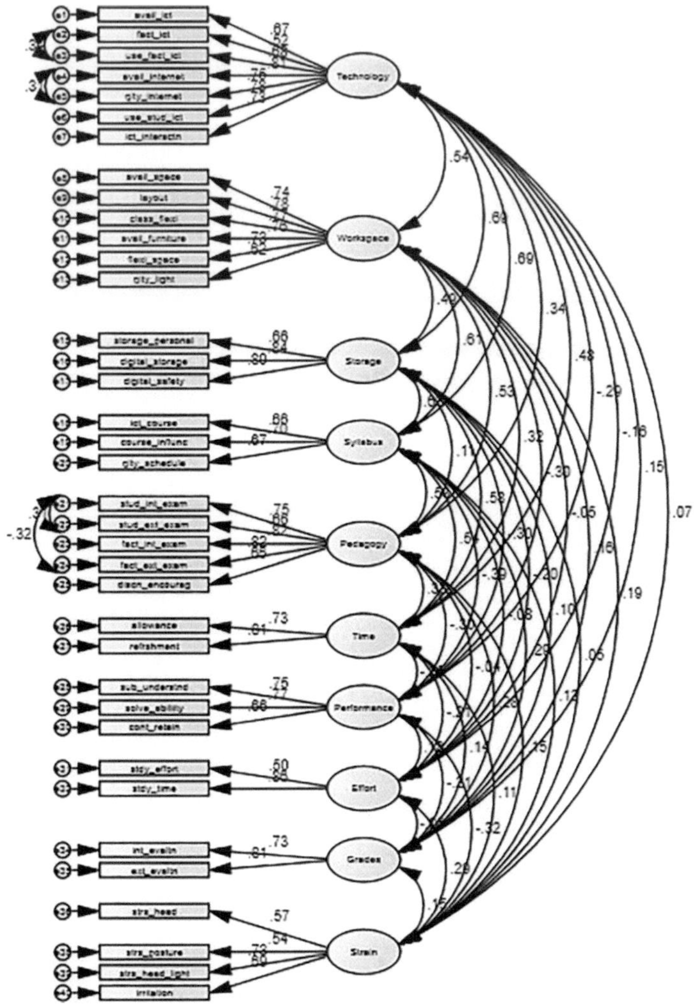

Fig. 14.5 Modified measurement model

Table 14.2 Model fit parameters of measurement model

Model	Chi-Sq. (df), p	RMR	GFI	CFI	RMSEA
Default	1588.467(695), 0.000	0.61	0.809	0.852	0.61
Modified	1216.68 (580), 0.000	0.59	0.839	0.917	0.56

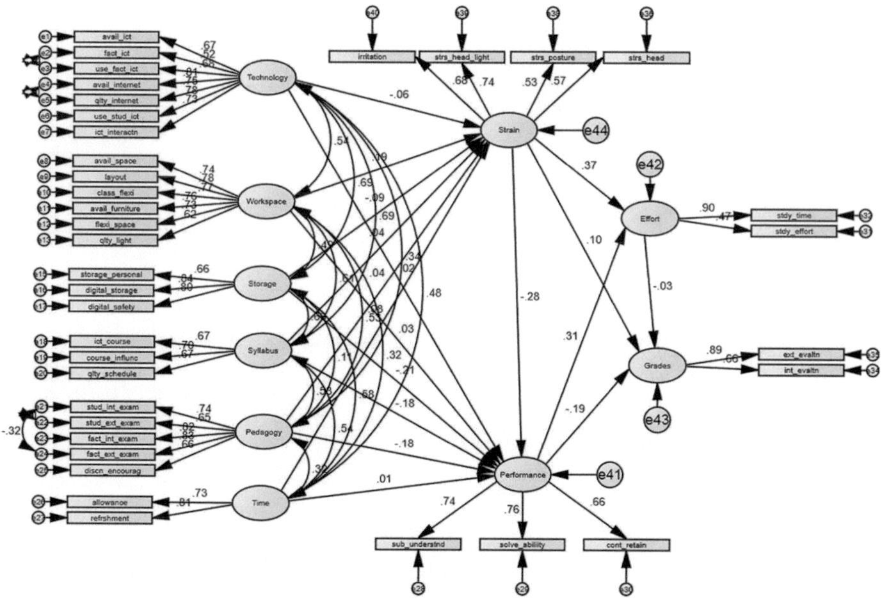

Fig. 14.6 Initial hybrid model for classroom model

Table 14.3 Modification history of the model for classroom worksystem

Model	Modification	χ2 (df), p	RMR	GFI	CFI	RMSEA
Default		1218.817 (582), 0.000	0.060	0.839	0.887	0.056
Change 1	Remove 'Technology', 'Performance' interaction, insignificant p-value	1233.878 (598), 0.000	0.061	0.837	0.887	0.055
Change 2	Remove 'Performance', 'Time' interaction, insignificant p-value	1235.324 (601), 0.000	0.061	0.837	0.888	0.055
Change 3	Remove 'Effort', 'Grade' interaction, insignificant p value	1236.959 (602), 0.000	0.059	0.837	0.907	0.051

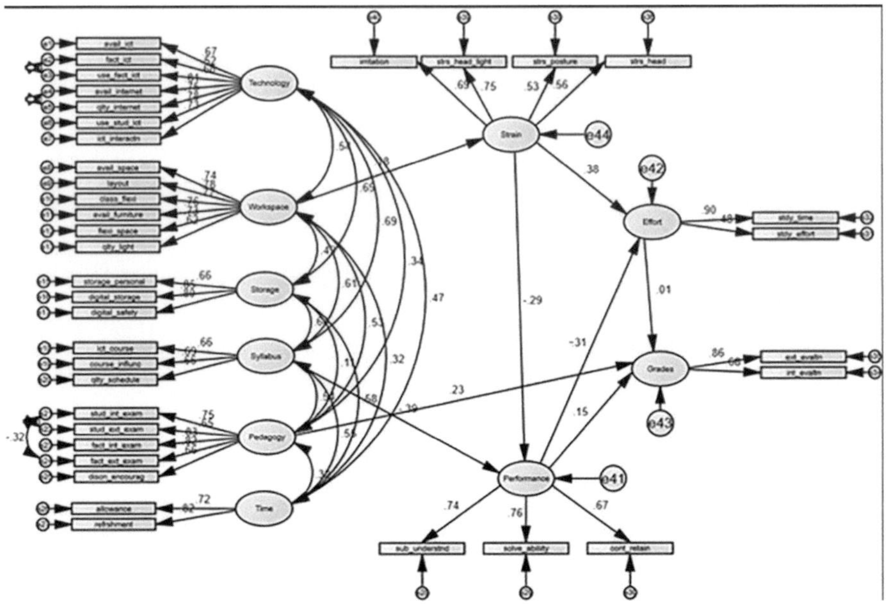

Fig. 14.7 Final structural equation model for classroom work system

14.5 Conclusion

A structural model has been developed representing the causal relationship between the classroom factors and outcome measures. The final structure has a model fitness parameters, $\chi 2 = 1236.95(df = 605)$, $p = 0.000$, $RMR = 0.059$, $GFI = 0.837$, $CFI = 0.907$ and $RMSEA = 0.051$. It has been understood that these classroom factors influences the outcome of the teaching–learning experience to a great extent.

Based on the study, a few recommendations have been made. Improvement of ‘Technology’ dimension could be made through better connectivity, Internet access to digital multimedia and display devices. Interventions to improve the use of ICT by faculty (Skill_ICT) need to be explored with respect to (i) availability and quality of ICT, (ii) teaching methodology or pedagogy before undertaking the interventions. Ways and means to improve workspace environment in class rooms need to be examined, i.e. through improvement of classroom configuration and/or physical environment systems or its design. It is suggested that a detailed anthropometric study be undertaken, followed by ergonomic design of the furniture and space configuration.

Acknowledgements The financial support provided through TEQIP Phase II to carry out expenses incurred during the project is acknowledged. Also the support provided by the entire faculty and student community of Rajiv Gandhi Institute of Technology, Kottayam, who spared their valuable time to participate in this work is acknowledged.

References

Beauchamp G, Kennewell S (2010) Interactivity in the classroom and its impact on learning. Comput Educ 54(3):759–766

Bridger RS (2008) Introduction to ergonomics. CRC Press

Conant RC, Ashby WR (1970) Every good regulator of a system must be a model of that system. Int J Syst Sci 1(2):89–97

Decker DM, Dona DP, Christenson SL (2007) Behaviourally at-risk African American students: the importance of student–teacher relationships for student outcomes. J Sch Psychol 45:83–109

Griggs MS, Gagnon SG, Huelsman TJ, Kidder-Ashley P, Ballard M (2009) Student–teacher relationships matter: moderating influences between temperament and preschool social competence. Psychol Schools 46:553–567

Karal H, Balicekapili T (2010) New perspective to educational technology: interdisciplinary cooperation -An example of faculties of education and engineering. Turk Online J Educ Tech 9 (1):132–142

McGrath Kevin F, Van Bergen Penny (2015) Who, when, why and to what end? Students at risk of negative student–teacher relationships and their outcomes. Educ Res Rev 14:1–17

Mohammed A (2013) Evolution of E-learning as a strategy of improving teaching and learning in nigerian universities: challenges and prospects. J Educ Pract 4(26):53–61

Murray C, Zvoch K (2011) Teacher-student relationships among behaviourally at-risk African American youth from low-income backgrounds: student perceptions, teacher perceptions, and socio-emotional adjustment correlates. J Emotional Behav Disord 19:41–54

O'bannon BW, Thomas K (2014) Teacher perceptions of using mobile phones in the classroom: age matters! Comput Educ 74: 15–25

Rajesh R, Reena M (2015) A review on worksystem interactions in a technology enabled class room. Int J Tech Enhanced Learn (in press)

Rajesh R, Reena M (2015) Structural equation modeling of classroom factors on outcome measures. International symposium in honor of Dr. A. Ravi Ravindran (OR2015), IISC Bangalore, India, March 12–13, 2015

Rajesh R, Revathy S, Nisha C (2015) A study on students' perception of current teaching-learning classroom environment. Indian Technology Congress, Bangalore

Rajesh R, Akhil CV, Muhammad NIP, Sachin VS, Vishnu K (2014) Survey on student's perception of class room environment in an engineering college. J Educ Pract 5(37): 80–91

Raju PLN, Gupta PK (2012) Satellite based education and training in remote sensing and geo-information: an e-learning approach to meet the growing demands in India. Int Arch PRSSIS 39:25–29

Author Index

A
Ashok, P., 49, 131

C
Chaitanya, V.V.V.N.K., 139

D
Das, B.B., 57
De, Jyotiraman, 103
Deshpande, Vishwas, 69
Drisya, S., 57
Durga Prasad, K.G., 139

H
Hima Gireesh, Ch., 139
Hridya, K., 163

I
Ikhar, Dhananjay, 69

J
Jeyapaul, R., 83

K
Karthick, J., 155
Krishnakumar, K., 117
Kumar Sharma, Harish, 37

M
Madhan Mohan, G., 49, 131
Maiti, J., 13, 25

N
Naveenkumar, R., 155
Nitish, K., 131

P
Parida, Ratri, 57
Prasad, M.V., 139
Princy, L., 163

R
Rajesh, R., 13, 25, 163
Raut, Abhijeet Arvind, 103
Ray, P.K., 1
Reena, M., 13

S
Saha, Esha, 1
Sakthi Nagaraj, T., 83
Satheeshkumar, M., 117
Shankar, S., 155
Singhal, Piyush, 37
Siva Sankaran, B., 49
Sonia, Pankaj, 37
Swapnesh, S., 163

T
Tewari, V.K., 1

P.K. Ray and J. Maiti (eds.), *Ergonomic Design of Products and Worksystems—21st Century Perspectives of Asia*, Managing the Asian Century,
https://doi.org/10.1007/978-981-10-5457-0

Subject Index

A
ANP, 139, 140, 142, 144, 146, 150, 151
Anthropometry, 107, 111, 114

B
Biomechanical, 25–29, 31, 34
Body mass index (BMI), 88

C
Chi-square test, 92
Cognitive, 158
Coir industry, 118
Combined manual material handling, 25–27, 29, 31, 34
Combined manual material handling task, 25
Cornell Musculoskeletal Discomfort Questionnaire (CMDQ), 89
Cube model-2, 13, 15

D
Data mining, 16
Decision tree, 13–15, 21, 23
Demographic and job characteristics, 88
Digital human modeling, 38
Discomfort survey, 87
DMQ, 69–71, 73
Drafting table, 139, 147, 150

E
Electromyography, 131, 134, 137
Ergonomics, 37, 106, 141, 143, 145, 150, 151

F
Fractional factorial experiment, 27

H
Heart rate monitor, 54, 131, 136
Human factor, 104, 105

I
Industrial workplace, 38
Intervention, 25, 26, 28, 29, 31, 33–35

J
Job stress, 158

K
Kano model analysis, 139, 140, 142, 143, 146, 150

L
Lean environment garment industry, 83
Lean manufacturing system, 86

M
Manual material handling, 13, 14
Mathematical model, 69, 70, 73, 80
Modeling and simulation, 37
MSDs, 57, 58, 60, 61, 65
Musculoskeletal disorder, 49, 50, 70

P
Physiological, 25–29, 31, 34
Physiological stress, 131, 137
Postural assessment, 57–59, 61, 63, 64
Posture analysis, 87
Prevalence, 118
Principal parameter(s), 2, 3, 6, 7
Product development, 111
Prolonged standing posture, 85
Prototype, 112, 113

Q
Quality function deployment, 139, 141

R
Radial drilling machine, 49–51

P.K. Ray and J. Maiti (eds.), *Ergonomic Design of Products and Worksystems—21st Century Perspectives of Asia*, Managing the Asian Century,
https://doi.org/10.1007/978-981-10-5457-0

Rapid Entire Body Assessment (REBA), 57–59, 61, 63–65, 90
REBA work sheet, 91
Repeated measures, 159
RULA, 57–59, 61, 64, 65, 131, 134, 137

S
Screen printing, 156
Sewing, 87
Sewing machine operators, 84
Standing modular sewing production, 98
Surface electromyography, 51, 53

T
Textile, 156
Textile industry, 84

U
U-shaped assembly systems, 85

W
WEKA, 13, 16, 17, 19, 20, 22, 23
Working postures, 131–134, 137
Work-related musculoskeletal disorders (WMSDs), 84, 117
Worksystems, 2, 5, 7, 8, 12

MIX
Papier aus verantwortungsvollen Quellen
Paper from responsible sources
FSC® C105338

If you have any concerns about our products,
you can contact us on
ProductSafety@springernature.com

In case Publisher is established outside the EU,
the EU authorized representative is:
Springer Nature Customer Service Center GmbH
Europaplatz 3, 69115 Heidelberg, Germany

Printed by Libri Plureos GmbH
in Hamburg, Germany